Patrick Moore's Pra

Other Titles in This Series

Navigating the Night Sky
How to Identify the Stars and Constellations
Guilherme de Almeida

Observing and Measuring Visual Double Stars
Bob Argyle (Ed.)

Observing Meteors, Comets, Supernovae and other transient Phenomena
Neil Bone

Human Vision and The Night Sky
How to Improve Your Observing Skills
Michael P. Borgia

How to Photograph the Moon and Planets with Your Digital Camera
Tony Buick

Practical Astrophotography
Jeffrey R. Charles

Pattern Asterisms
A New Way to Chart the Stars
John Chiravalle

Deep Sky Observing
The Astronomical Tourist
Steve R. Coe

Visual Astronomy in the Suburbs
A Guide to Spectacular Viewing
Antony Cooke

Visual Astronomy Under Dark Skies
A New Approach to Observing Deep Space
Antony Cooke

Real Astronomy with Small Telescopes
Step-by-Step Activities for Discovery
Michael K. Gainer

The Practical Astronomer's Deep-sky Companion
Jess K. Gilmour

Observing Variable Stars
Gerry A. Good

Observer's Guide to Stellar Evolution
The Birth, Life and Death of Stars
Mike Inglis

Field Guide to the Deep Sky Objects
Mike Inglis

Astronomy of the Milky Way
The Observer's Guide to the Southern/Northern Sky Parts 1 and 2
hardcover set
Mike Inglis

Astronomy of the Milky Way
Part 1: Observer's Guide to the Northern Sky
Mike Inglis

Astronomy of the Milky Way
Part 2: Observer's Guide to the Southern Sky
Mike Inglis

Observing Comets
Nick James and Gerald North

Telescopes and Techniques
An Introduction to Practical Astronomy
Chris Kitchin

Seeing Stars
The Night Sky Through Small Telescopes
Chris Kitchin and Robert W. Forrest

Photo-guide to the Constellations
A Self-Teaching Guide to Finding Your Way Around the Heavens
Chris Kitchin

Solar Observing Techniques
Chris Kitchin

How to Observe the Sun Safely
Lee Macdonald

The Sun in Eclipse
Sir Patrick Moore and Michael Maunder

Transit
When Planets Cross the Sun
Sir Patrick Moore and Michael Maunder

Light Pollution
Responses and Remedies
Bob Mizon

Astronomical Equipment for Amateurs
Martin Mobberley

The New Amateur Astronomer
Martin Mobberley

Lunar and Planetary Webcam User's Guide
Martin Mobberley

Choosing and Using a Schmidt-Cassegrain Telescope
A Guide to Commercial SCT's and Maksutovs
Rod Mollise

The Urban Astronomer's Guide
A Walking Tour of the Cosmos for City Sky Watchers
Rod Mollise

Astronomy with a Home Computer
Neale Monks

More Small Astronomical Observatories
Sir Patrick Moore (Ed.)

The Observer's Year
366 Nights in the Universe
Sir Patrick Moore (Ed.)

A Buyer's and User's Guide to Astronomical Telescopes and Binoculars
Jim Mullaney

Care of Astronomical Telescopes and Accessories
A Manual for the Astronomical Observer and Amateur Telescope Maker
M. Barlow Pepin

Creating and Enhancing Digital Astro Images
Grant Privett

The Deep-Sky Observer's Year
A Guide to Observing Deep-Sky Objects Throughout the Year
Grant Privett and Paul Parsons

Software and Data for Practical Astronomers
The Best of the Internet
David Ratledge

Digital Astrophotography: The State of the Art
David Ratledge (Ed.)

Spectroscopy: The Key to the Stars
Keith Robinson

CCD Astrophotography: High-Quality Imaging from the Suburbs
Adam Stuart

The NexStar User's Guide
Michael Swanson

Astronomy with Small Telescopes
Up to 5-inch, 125 mm
Stephen F. Tonkin (Ed.)

AstroFAQs
Questions Amateur Astronomers Frequently Ask
Stephen F. Tonkin

Binocular Astronomy
Stephen F. Tonkin

Practical Amateur Spectroscopy
Stephen F. Tonkin (Ed.)

Amateur Telescope Making
Stephen F. Tonkin (Ed.)

Using the Meade ETX
100 Objects You Can Really See with the Mighty ETX
Mike Weasner

Observing the Moon
Peter T. Wlasuk

Astronomical Sketching: A Step-by-Step Introduction

Richard Handy
David B. Moody
Jeremy Perez
Erika Rix
Sol Robbins

Springer

Richard Handy kraterkid@msn.com
David B. Moody bicparker@mac.com
Jeremy Perez beltofvenus@perezmedia.net
Erika Rix erika_rix@yahoo.com
Sol Robbins planetsketcher@aol.com

Cover illustrations courtesy of the authors.

Library of Congress Control Number: 2004934194

Patrick Moore's Practical Astronomy Series ISSN 1617-7185

ISBN-10: 0-387-26240-7   e-ISBN-10: 0-387-68696-7
ISBN-13: 978-0-387-26240-6 e-ISBN-13: 978-0-387-6896-7

© 2007 Springer Science+Business Media, LLC
All rights reserved. This work may not be translated or copied in whole or in part without the written permission of the publisher (Springer Science+Business Media, LLC, 223 Spring Street, New York, NY 10013, USA), except for brief excerpts in connection with reviews or scholarly analysis. Use in connection with any form of information storage and retrieval, electronic adaptation, computer software, or by similar or dissimilar methodology now known or hereafter developed is forbidden. The use in this publication of trade names, trademarks, service marks, and similar terms, even if they are not identified as such, is not to be taken as an expression of opinion as to whether or not they are subject to proprietary rights.

9 8 7 6 5 4 3 2 1   BS/EVB

springer.com

To the memory of my father, and to my dear friend, Henry L. Kline. Special thanks to Professors Robert Chiarito, Patricia Pepper, and Dr Joseph Wampler of UCSC. Your encouragement has kept me sketching and observing over the course of the decades.
—Richard H. Handy

To my wife and the memory of my grandmother, who both gave their unwavering support in all my writings and other endeavors.
—David B. Moody

For Amanda, Giselle, and Harrison. Thank you for your support and inspiration.
—Jeremy Perez

To my husband, Paul, for his endless love and support.
—Erika Rix

To Crystal and James, with their love and support all things are possible.
—Sol Robbins

Preface

From its very inception, this book was intended as an easy-to-follow introduction to sketching celestial objects. Every chapter presents several step-by-step tutorials with detailed sequential photographs and text. I believe it also reflects some of the wide diversity of media and techniques used to render astronomical objects. Bringing together four talented sketch artists was an expression of my desire that it sample the work of a community of astronomical sketchers, rather than being a compendium of an individual's sketches. Each co-author has a unique mastery of the media and brings a wealth of knowledge with a strong desire to share tips and techniques with the novice. Astronomical sketching is not only a powerful means for recording your visual observations—it is an exciting, personal relationship with the cosmos, a path of discovery, challenge, and experimentation. It is our fervent hope that you follow that path.

—Richard Handy

Acknowledgments

Richard Handy Working with such a talented group of astronomical sketchers has been a wonderful experience. I would like to thank Erika, Jeremy, Sol, and David for contributing so much time, energy, and devotion to every aspect of this unique collaboration. I am thankful for the relationships we have developed and the laughter that we share. To the broader community of amateur astronomical sketchers, your sketches are an inspiration and an endless source of delight.

Erika Rix There are so many amateur astronomers that have been an inspiration to me, willing to share their techniques, advice, and encouragement. I am very thankful for the opportunity to become a better observer and sketcher with their assistance. A special thank you is for Michael Rosolina, whose excellent sketches and reports laid the foundation of my fascination with Ol' Sol as well as Lady Luna. A personal thank you goes to John Crilly and Scott Kroeppler for their friendship and encouragement, as well the use of their books and equipment during my past studies of the Sun. Rich, Sol, Jeremy, and David, the four of you are amazing and it has been a joy to work so closely with you during the course of this book. Thank you Rich for allowing me to be a part of this experience.

Sol Robbins I would like to especially thank all other planetary sketchers who have selflessly shared with me their insights and talents through the Internet. Another personal thank you is also in order to Valery Deryuzhin, Al Misiuk, and Bill Burgess. Their optical knowledge and expertise enhanced my observing in helping me to turn my telescopes into great optical performers.

Acknowledgments

Jeremy Perez I would like to acknowledge Bill Ferris, whose advocacy and passion for astronomical sketching put me hot on the trail of this enjoyable facet of amateur astronomy. I also want to recognize the masterful work of Eric Graff, whose astronomical renderings taught me that deep sky sketches truly can reflect the beauty and detail visible through the eyepiece. To my fellow co-authors: it has been a rewarding and educational experience to collaborate with all of you in this effort.

David B. Moody People do not do things alone, even when they think they do. I have heard it said that " . . . behind every great man is a woman rolling her eyes." This is true in my case and I thank Beth more than she knows for rolling her eyes at just the right time and supporting me in this writing. I also have to thank Avery and Jones for their constant editorial attention and Bic Parker for his role as muse. It has been especially fun working with the talented authors in this book and I appreciate their allowing me to be part of this, especially Rich for inviting me, sight unseen. We often take the dark night skies for granted and I do not want to forget my appreciation for them. They are the ultimate inspiration for much of what you may read in this book.

Along the way, we have all received much inspiration from the fantastic community at CloudyNights.com where the discussion of all aspects of astronomical sketching finds a stimulating and enlightening home.

Finally, we all want to thank John Watson, Harry Blom, and Christopher Coughlin for their excellent advice and guidance. Without your support this book would not have been possible.

Contents

Preface vii
Acknowledgements ix
Introduction xiii

Chapter One   (Richard Handy)
Sketching the Moon 1
1.1 Graphite Pencil Sketching 2
1.2 Charcoal Sketching (by E.Rix) 6
1.3 Pen Sketching 10
1.4 White Chalk on Black Paper Sketching 16

Chapter Two   (Jeremy Perez)
Sketching Comets 23
2.1 Sketching a Comet and Its Motion 24
2.2 Creating a Wide Field Comet Sketch 32
2.3 Assessing Cardinal Directions 39
2.4 Sketch and Observational Log Sheets 40
2.5 Sketching Faint Objects in Low Light 42

Chapter Three   (Erika Rix)
Sketching the Sun 47
3.1 Basic White Light Sketching 50

3.2 Projection Sketching   52
3.3 Ha Filter Sketching: Prominences   58
3.4 Ha Full Disk Sketching   63

Chapter Four   (Sol Robbins)
Sketching the Planets   69
4.1 Sketching Saturn   72
4.2 Sketching Jupiter   78
4.3 Sketching Mars   86

Chapter Five   (Jeremy Perez)
Sketching Star Clusters   97
5.1 Sketching a Simple Open Cluster   99
5.2 Sketching a Complex Open Cluster with Unresolved Stars 105
5.3 Sketching a Simple Unresolved Globular Cluster 113
5.4 Sketching a Complex Globular Cluster 118
5.5 Marking Stars 126
5.6 Correcting Misplotted Stars 128
5.7 Stippling Technique 129

Chapter Six   (Jeremy Perez)
Sketching Nebulae 133
6.1 Sketching a Diffuse Nebula 134
6.2 Sketching a Planetary Nebula 141
6.3 Producing a Contour Sketch of a Dark Nebula 145
6.4 Producing a Shaded Sketch of a Dark Nebula 149
6.5 Using a Blending Stump 153
6.6 Using a Kneaded Eraser 157
6.7 Sketching Negative Versus Positive 159

Chapter Seven   (David B. Moody)
Sketching Galaxies 163
7.0 Getting Started 164
7.1 Your First Galaxy 166
7.2 The Next Step: An Irregular Galaxy and Dark Lanes 170
7.3 Tips and Tricks 172
7.4 Some Great Starter Galaxies to Sketch 175

Appendix A: Observing Forms and Sketch Templates 177
Appendix B: Glossary 183
Appendix C: Online Resources 189

Index 193

Introduction

Astronomical sketching has a long history in astronomy. Farmers before antiquity were sketching star patterns in stone, trying to determine the best time to plant their seeds. Galileo's first sketches of Jupiter showed the world how its moons made their orbits. Lord Rosse's sketches fired the imagination of Van Gogh. Barnard, known for his early photographs of the Milky Way, made volumes of sketches of the night sky for his research.

Why should we sketch the skies, especially in modern times? It might seem anachronistic when we have such incredible imaging tools at our disposal, capable of capturing such astonishing details. However, when you are involved in astronomical imaging with all of the required equipment, you will often find yourself immersed in an orchestra pit of computers, displays, and data. You will also spend much time aligning your equipment and calibrating your shots—all while a beautiful clear night sky rolls over your head in quiet majesty.

Some of you might simply want to look up and capture what your eyes see. Sketching allows you to capture what your eyes behold. There is some romance about seeing a planet or galaxy "live" and coming away with a piece of paper marked with evidence of what you saw. When you look back on these sketches months and years later, the marks you made on those sheets of paper will call to mind vivid memories of the celestial wonders you observed.

However, sketching does more than just create a memory book. It brings a number of advantages to the table for any observer. Sketching will train your eye to detect the features that an object possesses, especially the more elusive ones. We believe it is one of the best ways to help to develop and maintain an increased visual sensitivity. This higher acuity will help you to perceive subtle details at

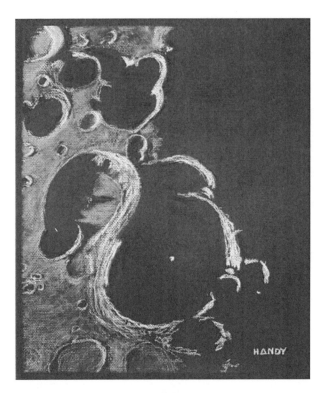

your threshold of vision. As you progress, you will undoubtedly feel a great sense of personal satisfaction. Whether or not you decide to share them with others, your sketches will serve as valuable records of your observations.

There have been arguments for and against sketching astronomical objects versus imaging them. Often, this debate hinges on the concern that there is undue subjectivity on the part of a visual observer. However, visual observing and imaging need not be mutually exclusive approaches. Instead, these methods can readily complement each other. The imager who sketches can see his images in

Introduction

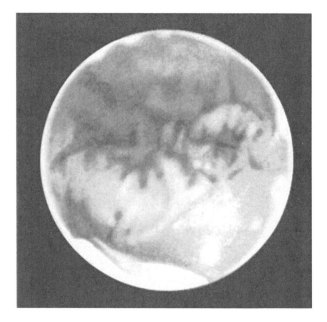

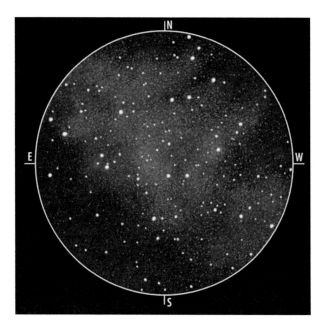

a context that he might not otherwise have without looking through the eyepiece. The sketcher who images can identify details in his images that he might have overlooked otherwise. Because these two ways of recording details can support and promote each other, we should not have to choose one to the exclusion of the other.

Introduction

Another thing to consider when sketching is that it is more important to record what was observed as faithfully as possible rather than trying to produce an artistic masterpiece. In time, you will notice that your drawings will improve and that the amount of detail you record becomes greater. Practice and persistence are the keys to becoming a better, self-trained observer.

An observer's personal quirks always add a unique character and personality to the sketches. Some of these quirks, however, might get in the way of a reasonably accurate record of the object being sketched. For example, one might have a tendency to draw belts upon Jupiter or Saturn too high or low or to crowd star patterns around a nebula and spread them too far apart in a cluster. In order to minimize any of these discrepancies, some standard techniques, drawing templates, and tips will be provided to you in each of the following tutorials. Use these tools and continue practicing your efforts, both at and away from the eyepiece. Given time, the details you record in your sketch will become very reliable.

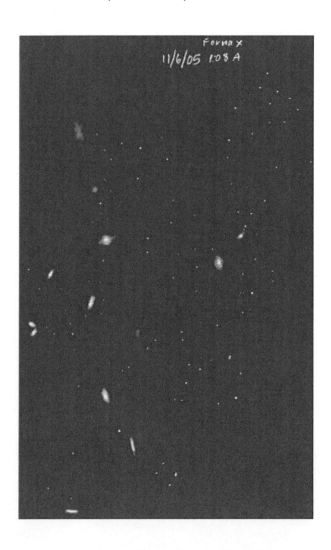

Introduction

The tutorials presented in this volume are meant to serve as a guide to get you started in this rewarding facet of amateur astronomy. You will be introduced to some of the media most commonly used to represent astronomical objects. Pencils, papers, ink, pastels, and charcoals are all discussed. Your choice of sketching medium and the technique you use might be based on the type of celestial object that you find the most intriguing or it might simply be a matter of personal preference. Feel free to experiment, try new approaches, and see what works for you. As you explore these techniques, keep in mind that just as each of the contributing authors has developed a unique style of sketching, so will you. In the end, it is all about an education of the eye, hand, and brain.

It is our hope that this work will inspire you to become not only a proficient sketcher but also a better observer.

CHAPTER ONE

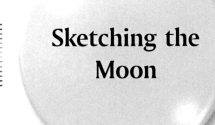

Sketching the Moon

If you have had the opportunity to observe the Earth's sister world through a small telescope or even a pair of binoculars, you probably already understand several aspects that make sketching her a delight. One of the most compelling is the excitement of creating a personal record of the astonishingly wide variety of terrains you observe. The Moon's face reveals a gold mine of impact- and volcanic-related processes: vast, basaltic lava-flooded basins, collapsed lava tubes snaking across its surface, long semicircular scarps that trace the shock from some of the most energetic collisions with ancient impactors, lofty mountain ranges that surround these basins, prominent isolated massifs, bright rays of pulverized rock that are flung hundreds of miles across the surface, and rugged highlands pitted by craters of all sizes. These are among some of the many treasures awaiting your discovery, and sketching these features is not only an education in observation, but it may also be a deeply rewarding personal record of the experience.

The intensely dramatic terminator—that imaginary line that traces the limits of the glancing rays of either the waxing or waning phase of the Moon—is perhaps the most challenging region to sketch. The light changes rapidly, making it difficult to render these transitory scenes; however its stark contrasts make forms stand out prominently, creating very alluring vistas. To make things easier to start though, you will probably want to try your hand at sketching features that are at least five degrees in lunar longitude from the terminator. The less tangent the light across the feature, the less obvious the changes will be during your sketching session. As your sketching skills improve, you will find that you can draw terminator features quickly and accurately.

Techniques Using Various Media

The first tutorial deals with drawing using graphite pencils, and then we will try our hand at sketching with charcoal. Following that, we will create a rendering using felt tip pen and end the chapter with an unconventional white chalk on black paper approach.

1.1 Graphite Pencil Sketching

Suggested art supplies

- Graphite pencils (I recommend using an H, B, 2B, and 4B to cover the typical range of lunar tones.)
- Erasers (Pink Pearl, Staedtler Mars plastic, Sanford Magic Rub, Art Gum or similar rubber or plastic erasers)
- Eraser shield
- Assorted blending stumps or tortillons
- Paper (I recommend 60–80-lb. acid-free Strathmore or Canson paper in 6″ × 9″ wire-bound pads.)

Preparation Graphite pencil sketches are generally done on white paper. You must be able to develop the gray tones over the entire area of the sketch quickly. Due to rapid changes in the angle of the Sun on lunar features, it is often best not to exceed 4″ × 5″ in sketching size. This size is ideal in fitting your sketch on a clipboard or to work from a small pad. In this tutorial, you can see the pad with a 4″ × 4″ outline that I have drawn. Next to it, I have indicated the pertinent data: date, sketch start and end times, lunar phase, seeing conditions (see the Glossary for a description of the Antoniadi scale), telescope type, focal length and aperture, eyepiece type and focal length, Barlow lens or diagonal (if used), Rukl Atlas page (or other lunar atlas pages), and, finally, a brief note of the name of the feature(s) that I am sketching.

I have selected a range of graphite pencils from hard lead "H", which produces a light shade of gray, to intermediate "B" and "2B" that produce gray tones, and the soft "4B" that can produce the darkest tones. You may find it possible to create very light to dark tones using a particular pencil (say a 2B) just by varying the pressure or angle of your stroke. Light pressure creates a lighter tone and a heavier pressure creates a darker one. I recommend practicing the widest possible range of gray tones on spare sheets of paper, using each pencil prior to your first graphite pencil lunar sketch.

Remember to use a comfortable observing chair and wear warm clothing if appropriate. Using a small flashlight helps to keep your sketchpad lit. If you are using a clipboard, you can purchase a battery-powered book light or an LED headlamp to illuminate your paper while you draw. The following steps require concentration and constant reference to the scene you view at the eyepiece. You can expect to spend anywhere from half an hour to an hour or more at the scope for each lunar sketch you make. Remember to indicate the start time before you begin your sketch.

Sketching the Moon

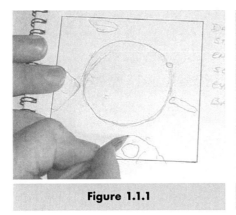

Figure 1.1.1

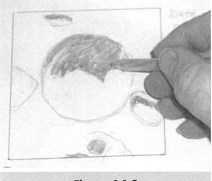

Figure 1.1.2

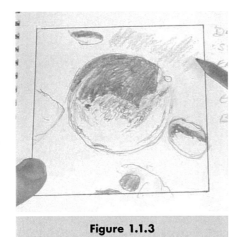

Figure 1.1.3

Step 1 Use the H pencil to create a light outline of the main features. Note that you do not need to be exact at this point, just delineate the size of main features to be included in the sketch and place them accurately in relation to each other. If you start with the largest crater, you can use it as a reference for the sizes and positions for the rest of the features. (Figure 1.1.1)

Step 2 Because the light changes rather quickly, it is a good idea to "rough in" the most prominent shadows at the beginning of the sketch using the 4B pencil. Do not worry about getting the precise shade yet, as you can establish the exact tone at a later stage in the sketch. Just indicate their positions and extent. (Figure 1.1.2)

Step 3 This is where the sketch starts to become really interesting. First, start adding detail to the crater rims and central peaks with the B pencil. Then, laying the pencil at a very shallow angle against the paper, use rapid, light, back-and-forth motions of the pencil to create a series of strokes on the page. The idea is to generate a light tone of gray that surrounds this group of craters. (Figure 1.1.3)

Step 4 Using the blending stump with moderate pressure, vigorously rub the marks, made in Step 3, somewhat randomly. Immediately you will see how this action softens the strokes, creating a uniformly even tone. At some point, the pencil strokes will disappear almost completely. If the tone is not quite dark enough, simply add similar strokes of the B or 2B pencil and continue blending.

Astronomical Sketching: A Step-by-Step Introduction

If it is too dark, erase to lightly remove excess graphite and then blend again. (**Figure 1.1.4**)

Step 5 Notice that the brightest areas of the sketch have been left untouched. This is something you will need to keep in mind as you add more tones to the sketch. By leaving the whitest tones free of graphite, you can effectively make use of the whiteness of the untouched paper to represent the brightest features, such as the crater rims and walls in this drawing. If you make a mistake, do not fret; use your eraser shield as shown to remove any undesired marks. (**Figure 1.1.5**)

Step 6 For the light gray tones on the crater floors, retrieve the H pencil and create tangent strokes as in Step 3, blending them as in Step 4. Be careful not to use the same stump or tortillon that you utilized for the darker areas unless you clean it with sand paper first, otherwise the result could be darker than you want. (**Figure 1.1.6**)

Step 7 Now you are reaching the point in the sketch where you can shift your attention to the shadows again, deepening them by applying strokes of the 4B pencil and then blending with the stump. As you do this, you can clearly see the obvious intensification of contrast in the drawing. (**Figure 1.1.7**)

Step 8 The sketch is nearly complete at this point. Make your final comparisons of the drawing to the image in your scope. Blend the areas to match your view through the eyepiece, adding any small details you may have overlooked. (**Figure 1.1.8**)

Step 9 Inevitably, your sketch will have smudges of graphite extending beyond the borders as this drawing clearly displays. Use your eraser shield

Figure 1.1.4

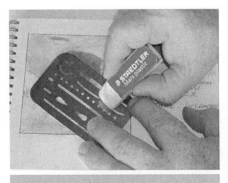

Figure 1.1.5

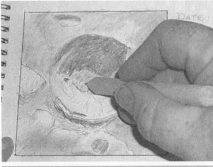

Figure 1.1.6

Sketching the Moon

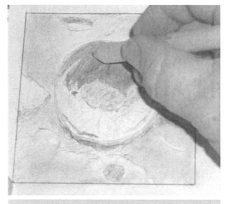

Figure 1.1.7

Figure 1.1.8

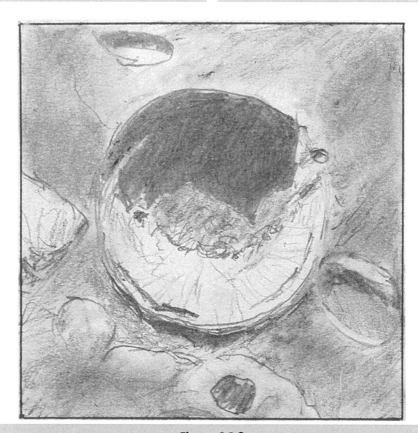

Figure 1.1.9

to mask the drawing from accidental erasure and cleanly remove the smudges with an eraser. At this point, the sketch is complete. Remember to record the ending time and any changes in seeing conditions that you noted during your session. (**Figure 1.1.9**)

Astronomical Sketching: A Step-by-Step Introduction

I would strongly recommend spraying the sketch with an acrylic fixative such as Krylon Clear Matte Finish no.7120. This will protect it from smearing or rubbing off on adjoining pages of your sketchpad. This is true for all the sketches presented in this chapter.

1.2 Charcoal Sketching (by E. Rix)

The whole concept behind this tutorial is sketching in layers. Charcoal is a wonderful medium, as it can portray the soft airy features as well as the most dramatic darker ones. Rarely will you require the use of erasers with this technique because if a mistake is made, it is easy to rub lighter or simply continue sketching over it. Highlighted areas have little or no charcoal and appear as the darker areas of the sketch take form.

Suggested art supplies

- $8\frac{1}{2}'' \times 11''$ sheet of sketching paper (I like the "Rite in the Rain" paper for nighttime sketching, as it is finely textured and has enough tooth to support the charcoal. Also, this type of paper is durable and weatherproof to withstand the effects that humid weather and dew have on regular paper, allowing you to continue sketching with little or no risk of your paper ripping apart due to dampness.)
- Two sizes of blending stumps
- Thick stick of charcoal
- Charcoal pencil in a wooden holder
- Sandpaper to sharpen your blending stumps
- Exacto Knife to sharpen your charcoal pencil
- Terry cloth rag to wipe your hands

Step 1 Study the feature closely and lightly sketch the outline of the crater with a sharpened charcoal pencil. Work in small strokes so adjustments can be made quickly if the shape is inaccurate. Shadow detail is immediately added for two reasons: (1) Shadows will change dramatically over a short period of time. By capturing the shadows immediately, their shapes will not be lost as the sketch progresses. (2) Placement of the features within the crater will be easier with the shadows assisting orientation. (**Figure 1.2.1**)

Step 2 Outlines of the interior wall of the crater can now be added with your charcoal pencil. This will be the foundation of the sketch. You will notice that a few smaller craters have been added outside of the main crater in **Figure 1.2.2**. Adding these now will help dramatically with the placement of other features later in the sketching process.

Step 3 Using a small blending stump that has been sharpened and cleaned with sandpaper, apply softer shadowing within the crater walls. Rub your blend-

Sketching the Moon

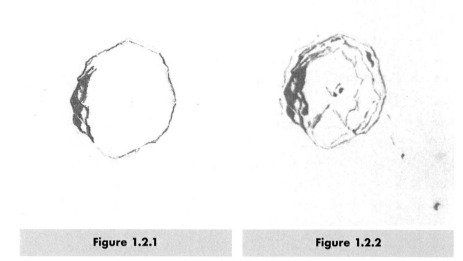

Figure 1.2.1

Figure 1.2.2

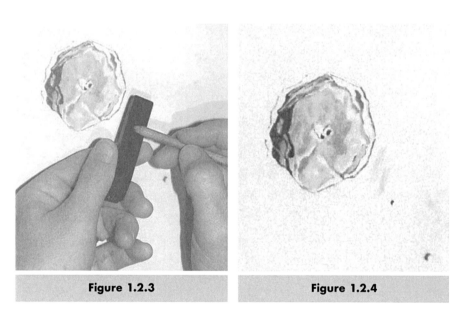

Figure 1.2.3

Figure 1.2.4

ing stump onto the stick of charcoal for this stage and apply the charcoal onto the paper with your stump. (**Figure 1.2.3**) You will notice how easily different depths of shading can be added with a stump, focusing on delicate flowing curves or sharper angles. Pay attention to the slightest details on the crater floor. The faintest rilles and ridges can be detected in this sketch. (**Figure. 1.2.4**)

Step 4 Once the background has been added to the inside of the crater, another layer of detail can be added over it with your charcoal pencil. Notice the addition

Astronomical Sketching: A Step-by-Step Introduction

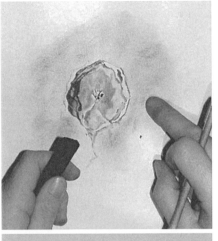

Figure 1.2.5

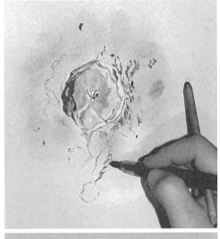

Figure 1.2.6

of the summits to the central mountain in **Figure 1.2.5**. The western wall is given more definition, as well as using a blending stump rubbed in the charcoal stick for the rugged ridge just outside the southern border of the main crater. At this stage, background for the entire sketch is added very softly by rubbing your fingertip onto the charcoal stick and applying it with light quick strokes.

Step 5 Add outlines of the features located outside the crater with a charcoal pencil or a stick of charcoal. Remember to outline the darkest striking features, not the highlighted ones. It does not matter if you make a mistake. Work quickly, letting your eyes soak in the full view, with your hand reflecting what your eyes see. (**Figure 1.2.6**) Take note of the terrain such as any nearby craters, wrinkle ridges, the ruggedness of steep tormented slopes, or any existing rays. The crater featured in this step is a complex crater with fascinating terrain. The goal at this stage is to capture the essence that makes this formation unique. (**Figure 1.2.7**)

Step 6 This stage of the sketch involves adding another layer to the lunar surface, setting the stage for the grand finale, so to speak. Use a large blending stump to smoothen harsh outlines. With textured paper, the difference between using your finger to blend, as compared to a stump, is quite remarkable. Using your finger to blend charcoal on textured paper will result in a courser blend. By using a soft stump, the appearance will be smoother. Notice the terrain starting to take form. You can almost visualize the finished result. (**Figure 1.2.8**)

Step 7 Add a slightly darker layer with charcoal, bringing definition to the background created. Use the charcoal pencil for finer lines and the stick of charcoal for broader additions. At this stage, the surrounding features start to take shape. (**Figure 1.2.9**)

Step 8 Now concentrate on the surrounding craters as well as the jagged ridgelines leading from the main crater to its companions. (**Figure 1.2.10**)

Sketching the Moon

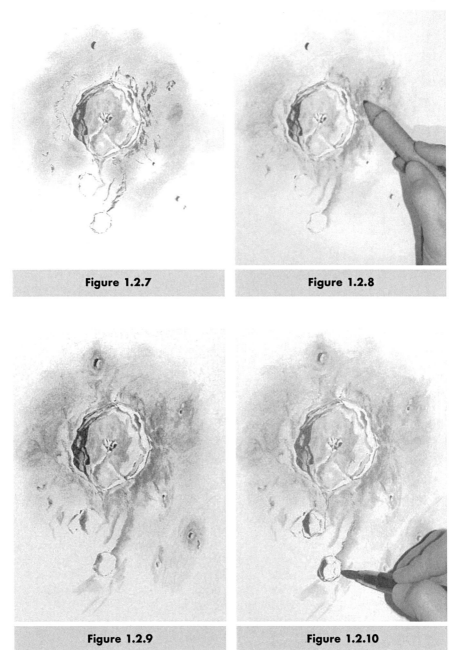

Figure 1.2.7

Figure 1.2.8

Figure 1.2.9

Figure 1.2.10

Step 9 Using a similar technique as with the main crater, apply shading inside the crater walls and their floors with a blending stump rubbed in charcoal. Additional shading is applied in the same manner on the lunar surface around them. The main crater has rays in this example, so no charcoal was added in

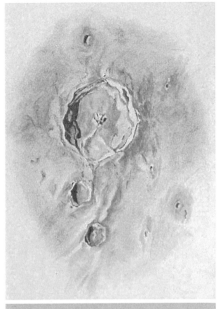
Figure 1.2.11

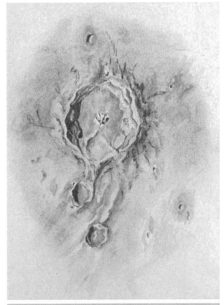
Figure 1.2.12

those areas. If too much charcoal has been added to these areas, rub it out with a clean blending stump, or if you prefer, a white vinyl eraser could also be used. (**Figure 1.2.11**)

Step 10 Nearly finished. One last layer of darker detail is added to any areas that require depth, vividness, and boldness. This step is done with a charcoal pencil. (**Figure 1.2.12**)

Step 11 Use a very fine blending stump to make the slightest adjustments so that vivid details are not overly softened. With this final touch up, your crater has completed its transformation onto the paper before you! (**Figure 1.2.13**)

Figure 1.2.14 was my first lunar sketch. Although this sketch was rendered using a 70-mm refractor compared to the 10″ reflector used in **Figure 1.2.13**, you can still see the dramatic improvement over a 9-month period. Aperture and seeing conditions played a part in the improvement; however, I believe that practice and improved observation skills were the key ingredients for an accurate recording of my night with Bullialdus in the tutorial sketch.

1.3 Pen Sketching

Traditional pen-and-ink sketching utilizes metal quills, plastic or wooden quill holders, and a bottle of India Ink in which the quill is dipped. Various-sized quills that will produce different line widths can be purchased at an art supply store. Drawing with this medium is not for the faint of heart, however, because you must be very careful not to put too much pressure on the quill or the ink will

Sketching the Moon

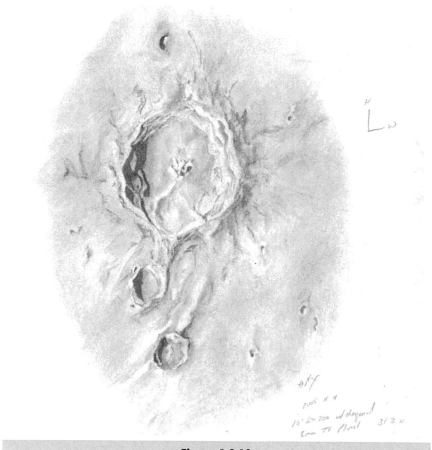

Figure 1.2.13

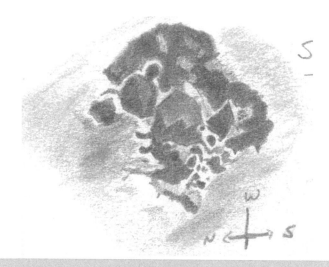

Figure 1.2.14

drip out onto the paper and create a large unintended blob. Ink sketching is relatively unforgiving of errors due to the simple fact that erasing ink is almost impossible to do without damaging your drawing. Although you can use "White Out" or something similar, the finished sketch will suffer some discoloration that will be somewhat distracting to the viewer. The stippling technique in this tutorial will not use these traditional methods. Instead, I will present a felt-tipped pen technique that is analogous to it.

Suggested art supplies

- Felt-tipped pens
  Sakura Color Products Corporation of Japan manufactures some wonderful archival quality, hard felt-tipped pens that are available in a wide range of precision tip diameters measured in fractions of a millimeter. These acid-free pens are called "Pigma Micron" and are inexpensive enough that you can purchase a number of them with different tip sizes for various stippling effects. I recommend buying the 0.20-mm, 0.45-mm, and 0.50-mm sizes. For broader areas of solid black shadow, try purchasing a "Writer" MS-600, this is an acid-free, archival-quality black marker that is part of the "ZIG memory system" manufactured by Kuretake Co., LTD. of Japan. You can also buy wider chisel point markers from Kuretake, although these may not be a necessity, given that these stippled drawings are smaller in size.
- Paper
  A small sketchpad of Strathmore or Canson, 60–80-lb. acid-free paper no larger than 6″ × 8″ is fine. I believe that a stippled sketch should not exceed 3″ × 4″ because of the amount of work required to create a drawing using the stippling technique. As an example, I did a 10″ × 10″ ink stipple drawing that took almost 15 hours to complete. Needless to say, I will not be working at that scale again anytime soon!
- Pencil
  You will need a hard H and 2H pencil to create the outline of the features that you observe that will be erased entirely once the pen work is complete and has dried.

Preparation Stippling requires producing randomly positioned dots of black ink so that spacing density will be perceived as various gray tones to the eye. A good way to get a feel of how this works is to draw ten 1/2″ × 1″ rectangles or equivalent sized squares as shown in **Figure 1.3.1**. Number them left to right, 1–10 underneath each rectangle. Note that the first rectangle on the left is the pure white of the page and the gray tone becomes darker the further to the right you go until the last rectangle, which is pure black. In this sketch, I used just the Pigma Micron 0.45-mm felt-tipped Pen. Although the dot size is the same throughout squares 2–9, the increased density gives the impression of differing and distinct gray tone values. Note the numbering of the rectangles as well. This is important, because as you create your outline sketch, you will code the various lunar features with these numbers and fill in the area with the density of dots that is associated with that respective rectangle. I usually attach this code sequence chart adjacent to my sketchpad on a clipboard. That way, while I am observing, I can continually make reference to the code number appropriate to that area of the sketch.

Sketching the Moon

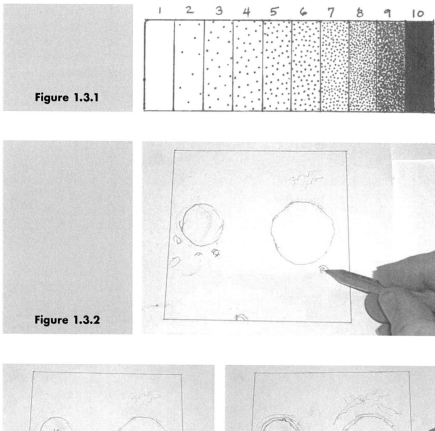

Figure 1.3.1

Figure 1.3.2

Figure 1.3.3

Figure 1.3.4

Step 1 Start by lightly outlining the rough shape of the main crater with the H pencil. At this point, do not worry about adding code numbers. (**Figure 1.3.2**)

Step 2 Outline the deepest black shadows and use the 2H pencil to draw a line connecting them to the blank space off to the side of the drawing area. Label these #10. (**Figure 1.3.3**)

Step 3 In similar fashion to Step 2, draw the outline of the brightest white areas and connect a line to the blank space adjacent to the sketch area. Label these #1. (**Figure 1.3.4**)

Astronomical Sketching: A Step-by-Step Introduction

Step 4 Repeat this procedure from the darkest gray tone (#9) to the lightest gray tone (#2), outlining and labeling until you have finished delineating every area of the crater and its included surroundings. The result will look very much like a "Paint-by-Numbers" sketch. You are now ready to step away from the telescope and return to a table inside, where you can finish the sketch in relative comfort and without the pressures of changing light. (**Figure 1.3.5**)

Step 5 Fill in the deepest black shadows with the black "Writer" marker, making sure you stay within the confines of your outline to avoid exaggerating the length or shape of the shadows. (**Figure 1.3.6**)

Step 6 Start with the lightest tone of gray (#2) and fill in the outline, duplicating the dot density in square #2 of the code sequence. Go on to fill in the next darker gray tone (#3), again by matching the dot density of the #3 square. Continue doing this until all of the gray tones from #2 to #9 are filled in with their correct values from the code sequence chart. (**Figure 1.3.7**)

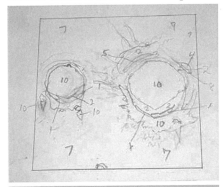

Figure 1.3.5

Step 7 Now that all of the outlines are filled in and have had time to completely dry, the last procedure is to erase the light pencil outline, revealing the completed sketch. Try not to be overaggressive with the pressure of the eraser; too much force could rip the paper. Art Gum erasers work well for this last operation because they are soft and crumble easily. (**Figure 1.3.8**)

Step 9 This is the completed sketch. (**Figure 1.3.9**)

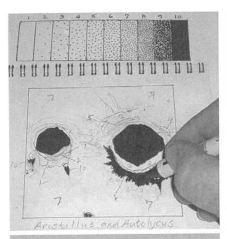

Figure 1.3.6

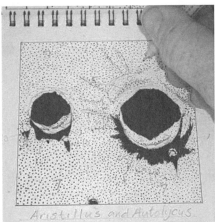

Figure 1.3.7

Sketching the Moon

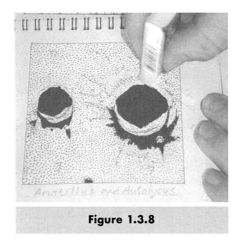

Figure 1.3.8

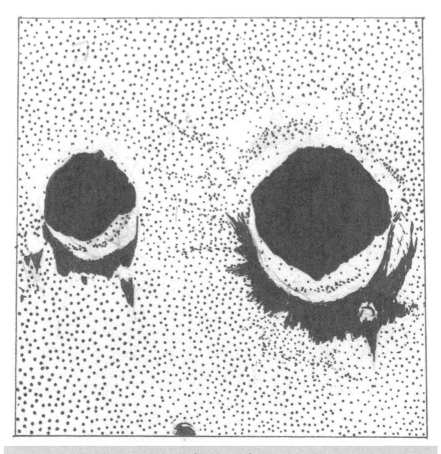

Figure 1.3.9

1.4 White Chalk on Black Paper Sketching

The white chalk on black paper medium is unique for the sketch development speed and realism that it can impart to a lunar drawing. Along with these wonderful attributes, the novelty of this approach is very attractive to those who like to explore new media. The dark shadows are created by what remains untouched by the white chalk. Consequently, only the bright whites and gray tones need to be developed. This is quite an easy thing to do with drawing chalks, because gray tones are produced by rubbing the chalk over the surface of the paper and blending it with a stump, tortillon, or sponge. The directly applied chalk creates the brightest white areas. As a result of the rapidness of the drawing progression, white-on-black sketches can be drawn at much larger scales than most traditional techniques. This allows the lunar sketcher the opportunity to draw large areas of the Moon's surface, yet still have the time to give the details adequate attention.

Suggested art supplies

- Conté Crayons (chalks)
  Conté Crayons are a fine French sketching chalk that have remarkable blending characteristics. From a dense layer to an extremely light gradation, they can be applied, smoothed, and manipulated to represent any feature or terrain. They come in compressed sticks and pencils of several colors, but for the purposes of creating a lunar drawing, the Blanc (white) HB and 2B as well as the Noir (black) B and 2B are ideal.
- Paper
  Purchase several 19" × 25.5" sheets of Strathmore textured black paper, Item # 107–110. This paper has amazing tooth—the textured quality of a paper's surface related to its ability to "hold" pigments or powders. Another great attribute is its thickness, 0.013" thick (fully 13 times thicker than computer paper!) Consequently, in very moist dewy conditions, this paper holds up wonderfully.
- Erasers
  I like the Staedtler Mars plastic eraser, the Paper-Mate Pink Pearl, and The Sanford Magic Rub.
- Blending stumps, tortillons and sponges
- Sanding block
  An inexpensive sanding block with several layers of sandpaper provides a way to sharpen the Conté crayons.
- Carpentry pencil sharpener
  These plastic sharpeners can be purchased at hardware or lumber supply stores. They have wide openings that allow the flattened lead carpentry pencils to be easily sharpened to a point. I have noticed they are great for sharpening the Conté Crayon sticks as well.
- Cosmetic brush

Sketching the Moon

Figure 1.4.1

Figure 1.4.2

- Masking Tape (2" wide)
- Sketch board
  If you use the full sheet of Strathmore paper as I do, a sketch board made of 1/8"-thick 24" × 30" Masonite, or press board, will allow you to position the paper and tape it down with room to spare for a hand hold. Obviously, you will probably want to make or purchase a smaller sketch board if you use half or quarter sheets.
- Easel
  An inexpensive wooden easel will make it easy to hold the sketch board while drawing. I made an easel out of a stepladder, which works quite well for me. It is a metal ladder that has plastic steps and the last piece at the top is a tray. I drilled a couple of 1/2" diameter holes and cut two 3" lengths of wooden dowels of the same diameter. The dowels were then inserted into the tray holes. The metal bar that arcs over the top of the ladder forms the backing for the sketch board. I put my drawing kit in a cardboard tray taped to the top step. (Figure 1.4.1)

Preparation Tape the paper to the drawing pad as shown in **Figure 1.4.2**. Note that half the masking tape's 2" width overlaps the paper edges. This will securely hold the paper down to the sketch board as well as provide a pleasing

Astronomical Sketching: A Step-by-Step Introduction

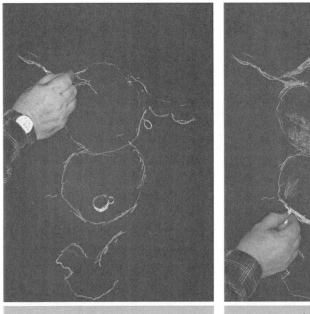
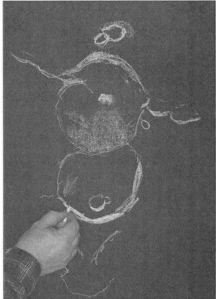

Figure 1.4.3   Figure 1.4.4

black border around the periphery of the drawing after the tape is removed. You can also use this taped border to jot down your sketch data: date, time session starts and ends, telescope size, and so forth. Just remember to transfer the data to a log entry sheet or write it on the back of the sketch so you do not lose it when the tape is peeled off. Place the drawing board close to you on your easel if you are using a full sheet, otherwise you may prefer to hold it during the sketch.

Step 1 Sharpen your HB white Conté Crayon chalk to a rounded point and then using the chalk, lightly outline the main craters' shapes, sizes, and relative positions, starting with the largest crater, using the HB white Conté Crayon. As in the previous tutorials, do not concern yourself with getting the exact contour; this level of detail will develop in the later steps. (**Figure 1.4.3**)

Step 2 Observe where the brightest whites occur in, and around, the lunar features. Add these to the rough outline with the white 2B Conté, paying close attention to their extent and positions. Here you can see the brightly lit inner rim of the main crater taking form. It is at this step that you should try to add detail to the rough outline. Make generous use of the erasers to adjust the main feature contours until you are satisfied that they look like the features you are observing. (Figure 1.4.4)

Step 3 Deep shadows are treated differently in this medium than the others previously discussed. Create a light outline around the shadows as shown in

Sketching the Moon

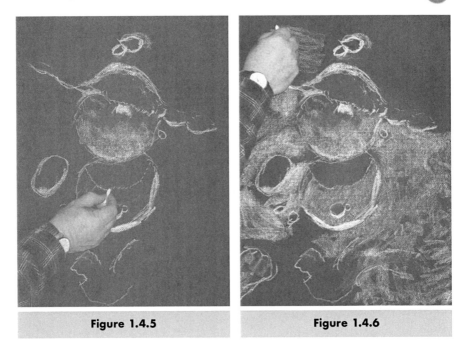

Figure 1.4.5

Figure 1.4.6

Figure 1.4.5. You will note that where the jagged edges are pierced with shafts of light, as they are on this large crater's floor, you can draw lines to indicate their presence. If a mistake is made and you extend the line either too long or wide, you can erase back to the blackness of the paper. There is no need to worry about getting the paper completely clean of the white Conté Crayon, however, because you can correct this later by adding black Conté Crayon over it if needed.

Step 4 The gray tone values are added at this stage of the sketch. Hold the 2B Conté stick on its side and lightly, but rapidly, in a back-and-forth random motion, build up the surface with strokes of white. Make certain that you do not cover the outlined shadow areas already established in the previous step. (Figure 1.4.6)

Step 5 Blend the white that you have overlaid on the surface of the paper with a natural or synthetic sponge. Rapid and randomized motions seem to work best, creating a uniform smooth gray tone. If you feel that the tone is too light, simply erase the area with an Art Gum eraser, clean the eraser crumbs off with your brush, and use the sponge without applying any chalk. This will allow you to produce a slightly darker gray value, using any residual chalk that has built up on the sponge. Pay attention to the gray tones of the lunar surfaces seen at the

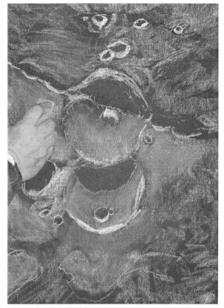

Figure 1.4.7

Figure 1.4.8

eyepiece and try to duplicate them as accurately as possible in your sketch. (**Figure 1.4.7**)

Step 6 Touch up the crater contours with the sharpened white 2B Conté Crayon. Constantly make reference to the shape of the rims, terraces, central peaks, rilles, craterlets, and other prominent features. (**Figure 1.4.8**)

Step 7 Finally, touch up the shadows that may have residual white markings from errant strokes or smudges that you cannot readily erase. You should use a sharpened, soft 2B black Conté Crayon, carefully stroking this area, taking care to avoid overextending your black layer. This is very important because white Conté Crayon does not layer well on top of black Conté. By making vigorous small back-and-forth motions with a clean sponge, large

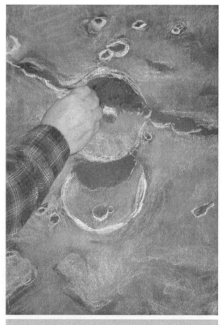

Figure 1.4.9

Sketching the Moon

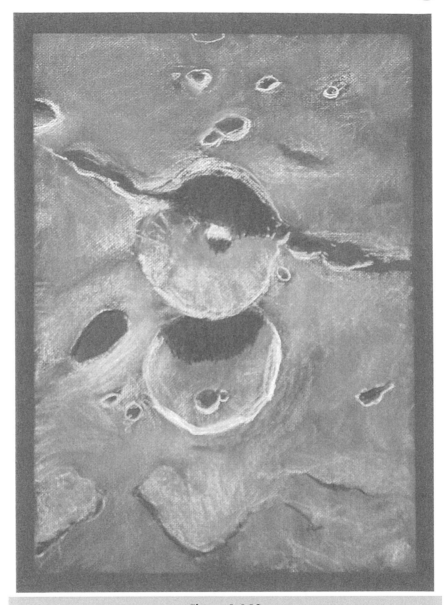

Figure 1.4.10

shadowed areas can be smoothed to produce very dark, even shadows that will blend with the blackness of the paper. (**Figure 1.4.9**)

Step 8 Carefully remove the masking tape from the periphery of the drawing. The sketch is now finished and this is the final result. (**Figure 1.4.10**)

CHAPTER TWO

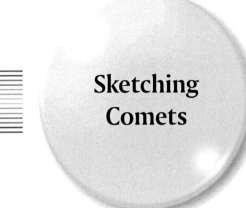

Sketching Comets

Comets add a sense of surprise and freshness to the predictability and seeming timelessness of the visible cosmos. Some of these mists of dust and fluorescing gas sail through the inner solar system at regular intervals, such as the famous comet 1P/Halley. Many other comets are discovered yearly as they make their first observed descent to our vicinity. Depending on their distance, composition, and intrinsic brightness, comets can present a variety of appearances—from almost stellar objects, to soft round patches, to majestic, tailed plumes that are sometimes visible to the naked eye. Because these are fleeting, transitory objects, time spent observing and sketching them is all the more precious.

Since comets travel very quickly as they approach the neighborhood of the Sun, it is often possible to note the gradual motion of the comet across the star field while observing telescopically. Although this can make locating a particularly faint comet difficult, this motion is something you may wish to record in your sketch. In the tutorials that follow, we will take a look at techniques used to capture the details that each comet possesses. Not only will it serve as a record of your experience, but it will also improve your observing skills and provide you with a great resource if you wish to discuss your observation with other comet enthusiasts.

The tools you will need to sketch a comet can be as simple as a sheet of paper, a pencil, a clipboard, and a red light. However, to give you more control over the appearance of your sketch, there are a few other items you may want to have ready. Following is a list of suggested sketching materials for a graphite comet sketch:

- Clipboard
- Dimmable red observing light
- Paper prepared in any of the following ways:
  - Blank
  - Prepared with predrawn sketching circles
  - Copied or preprinted log sheets
  - Copied, printed, or traced star fields (as discussed in the tutorials on pages 32 and 145)
- HB and 2H pencils
- Pen (for notes)
- Blending stump or tortillon
- Choice of erasers (Art Gum, eraser pencil, kneaded)
- Eraser shield (used to constrain erasures to a small area)
- Pencil sharpener or lead pointer
- Sandpaper block (used to hone the point of a pencil, blending stump, or tortillon)
- Small paint brush (used to brush away loose graphite or eraser debris)

2.1 Sketching a Comet and Its Motion

I will use a telescopic sketch of comet 73P-C/Schwassmann-Wachmann 3 as an example in this tutorial. This comet was relatively bright at the time of the observation at about 6th magnitude. Its beautiful, soft glow swept across the field of stars like the expanding wake of a speed boat on glassy water. The pseudonucleus was stellar in appearance with an elongated bright spike extending to the southwest. The flowing tail was striated with varying levels of luminosity and consisted of a bright, narrow, inner portion enveloped by a shorter, fainter, outer fan.

When observing comets, use averted vision, different eyepieces, and even slight movements of the telescope to discern as many details as possible. Take nothing for granted. Fascinating unexpected structures may reveal themselves, if you are diligent with your observation and sketch. Not all comets display bright tails, and all you may see at first is a round coma. As you observe this feature, try to discern how strongly defined its core is. At higher magnification, see if you can you discern any subtle irregularities in brightness.

If the comet is faint, a tail (if present) may be very difficult to detect. This is where movement of the telescope—perhaps letting the comet drift through the field—can be helpful in spotting this feature. If one or more tails are evident, how long and wide do they appear? Which portions of the tails are brightest, and can you spot any irregularity in that brightness? The comet in this tutorial exhibits a pronounced tail. If the comet you are sketching does not have one, simply disregard those steps.

Choose a magnification that you feel provides a good view of the comet. You may decide to render more than one sketch to cover details seen at low and high magnifications. Or you may combine multiple views into one sketch. For this sample sketch, I chose a low-power view that nicely highlighted the comet's bright

Sketching Comets

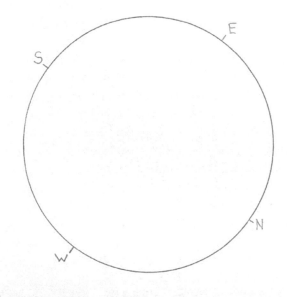

Figure 2.1.1

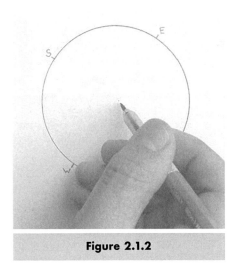

Figure 2.1.2

tail. I made the observation using my 15-cm f/8 Newtonian with a 32-mm Plössl eyepiece. This offered a magnification of 37.5× and true field of view of 88 arc minutes.

Step 1: Framing and preparing the sketch area Line up your telescope so that you get the best view of the comet and its surroundings. If you have some room for adjustment, try to place a conspicuous star in the center. This is not absolutely necessary, but it can make the placement of stars and comet features more convenient. Mark the cardinal directions around your sketch circle. (**Figure 2.1.1**) Use one of the techniques discussed in *Assessing Cardinal Directions* on page 39.

Step 2: Plotting the star framework If you placed a star in the center of your view, mark it now with your pencil. (**Figure 2.1.2**) For tips on plotting stars in your sketch, see *Marking Stars* on page 126. If you have placed the core

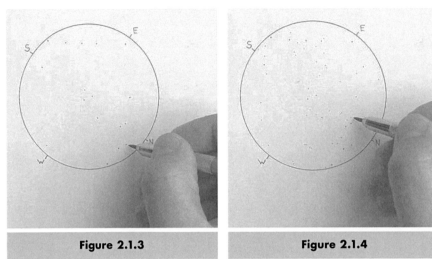

| Figure 2.1.3 | Figure 2.1.4 |

of the comet at the center of your sketch, refer momentarily to Step 3 to lightly mark its position there and then return to this step.

With the center point marked, you will now plot the remaining stars that are visible through the eyepiece. (**Figure 2.1.3**) These stars will serve as measuring points when you add dimension and note the motion for the comet. They will also allow you to describe the position of the comet in relation to known star positions using a star atlas or planetarium software. I recommend going through the *Sketching a Simple Open Cluster* tutorial on page 99 for tips on how to accurately place these stars. Simply put, plot the brightest stars first by imagining their positions on a clock face, noting how far they are from the center. Then proceed to mark the fainter stars by either using the same method or by noting where they reside in relation to the stars you have already plotted. Mark these stars lightly, so there is a notable difference between the bright and faint stars. Try to visualize the geometric shapes they make with each other as you do this. Finish this stage by comparing the boldness of the stars in your sketch with the eyepiece, making progressively brighter stars bolder if necessary. (**Figure 2.1.4**)

Step 3: Marking the comet's position With the framework of stars in place, you can now sketch the comet. To begin, I suggest using a blending stump to lightly mark the center of the coma. Refer to the techniques described in *Using a Blending Stump* on page 153 to lightly load your blending stump with graphite. With the blending stump prepared, lightly swirl a very small mark at the comet's position (**Figure 2.1.5**). As soon as you do this, check the time and mark it in your notes.

NOTE: In this tutorial, I render the tail first, since it was so prominent. However, if the coma of the comet is more pronounced, you may wish to sketch it first (see Step 6). The order in which these features are sketched should not be set in stone. In fact, you will likely switch back and forth between them as more details reveal themselves over the course of the observation. For simplicity's sake though, we will tackle them one at a time.

Sketching Comets

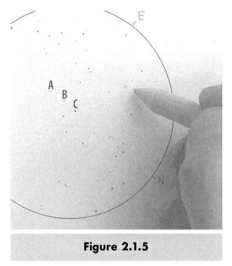

Figure 2.1.5

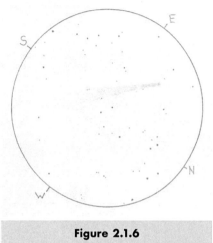

Figure 2.1.6

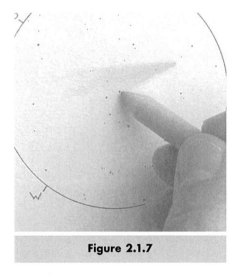

Figure 2.1.7

Step 4: Shading the core of the tail If the comet has two or more widely separated tails, concentrate on rendering one at a time. Rendering the bright core of the tail first will give you a spine on which to build the rest of the tail. Using the stars in your sketch as a guide, note which direction the tail points. Use averted vision to determine how long it appears and how wide it is. Load your blending stump and begin with the portion of the tail that is the most obvious. For this sketch, the bright, inner core of the tail was spread between two stars marked A and C in **Figure 2.1.5**. There was also a brighter spike within: that pointed to the star marked B in the same image. The tail as a whole extended past the field stop of the view, but was rather faint by that point.

To render the tail, lightly load your blending stump with graphite. Beginning at the head of the comet, use circular, elliptical, and even linear strokes to define the core of the tail. (**Figure 2.1.6**) Lighten pressure on the blending stump as you render fainter areas. Keep your eyes open for clumps or streamers of brightness so that you can render these faithfully in your sketch. (**Figure 2.1.7**)

Step 5: Refining the tail In order to see the fainter portions of the tail, you will need to give your eyes a rest from your red sketching light. Spend plenty of time at the eyepiece, using averted vision and slight motions of the telescope to bring out as many faint details as possible. With the image sufficiently burned in your mind, load your blending stump with a small amount of graphite. Then

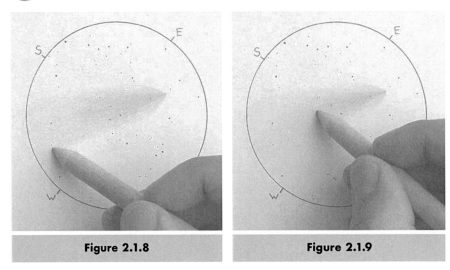

| Figure 2.1.8 | Figure 2.1.9 |

use soft, smooth strokes to render these fainter portions of the tail. (**Figures 2.1.8**) To produce a gradual transition at the outermost edges of the tail, reduce the pressure on the blending stump to almost nothing in those areas. Return to any brighter portions of the tail and darken them up if necessary. (**Figure 2.1.9**) If you find that you have shaded an area too heavily or produced a transition that is too sharp, you can use a kneaded eraser to carefully lift graphite away from these areas (see *Using a Kneaded Eraser* on page 157).

Step 6: Adding the coma Using other stars in the view for comparison, determine how wide the coma appears. With some comets, the coma will appear circular and distinct from the tail. Other comets may not offer such a simple distinction. For example, in this sketch of 73P-C/Schwassmann-Wachmann 3, the coma did not present itself as a distinctly circular feature. There was, however, a faint outer shell to the tail near the comet's head that will serve to illustrate the point.

To render this feature, load your blending stump very lightly with more graphite. Using a very soft, circular, or elliptical motion, begin at the center of the coma and swirl outward, lightly defining its boundaries. (**Figure 2.1.10**) If the outer edges of the coma are soft, reduce pressure on the stump to almost nothing as you shade these areas. (**Figure 2.1.11**) Observe how bright the coma gets as you approach its core and add more layers of graphite with your blending stump to represent this.

Be sure to capture any luminous structure noted in the coma, such as jets, hoods, and fountains, with increased pressure on your blending stump. Dark areas in the coma can be captured by gently removing graphite with your kneaded eraser. Pay close attention to how the scale and position of these features relate to the overall size of the nucleus. Using clock face and concentric circle imagery can help here as well.

Step 7: Adding a pseudonucleus or central condensation If the comet presents a bright, stellar, or almost-stellar point at its heart, determine

Sketching Comets

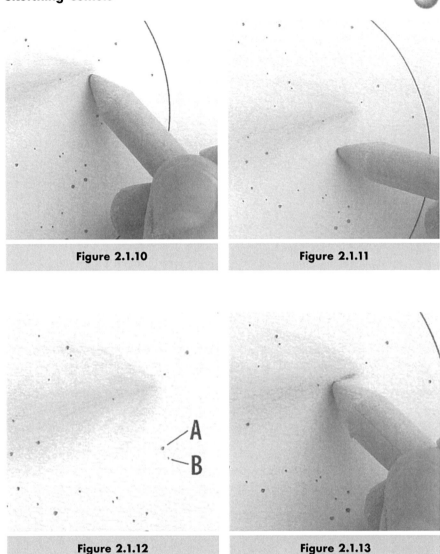

Figure 2.1.10

Figure 2.1.11

Figure 2.1.12

Figure 2.1.13

how sharp it is. If it is a thick, bold spot or streak, you may want to use your blending stump to mark it. Pick up more graphite on the stump if necessary and then carefully press it at the position of the central condensation. Apply controlled pressure with a very tight circular motion until the correct brightness is achieved. If there is a stellar pseudonucleus at the core, use your pencil to carefully mark it. Compare it to any other stars in the field that are similar in brightness and mark it with a similar boldness.

In this example, the pseudonucleus was stellar in appearance, but had a bright streak extending away from it in the direction of the tail. This streak was about as long as the distance between the stars marked A and B in **Figure 2.1.12. Figures 2.1.13**

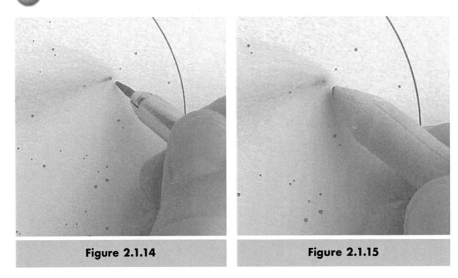

Figure 2.1.14   Figure 2.1.15

and 14 demonstrate the addition of this streak and the plotting of the stellar pseudonucleus. If this point appears to be too sharply defined, further blending may be needed. To do this, lighten the load of graphite on your blending stump, return to the core of the coma, and lightly swirl outward to soften the transition. (**Figure 2.1.15**) If this lightens the pseudonucleus too much, you will need to reapply more graphite. Repeat these steps until you are satisfied with its appearance.

Step 8: Finishing the sketch Take an overall look at the comet now and compare it to your sketch. Rework any inconsistencies with your blending stump and kneaded eraser. After shading the comet features, replot any stars that were blurred in the process. Take a few moments to reexamine the view through the eyepiece and compare the star brightness to what you see on your sketch. Adjust the boldness of any stars necessary to match what you see. (**Figure 2.1.16**) This is also a good time to finish writing any notes about the observation.

Step 9: Marking motion I find that part of the enjoyment of comet observing is noting the motion of the comet over the course of the observation. You may notice motion over the course of making your sketch or you may go off to other observations and return to the comet an hour or two later to note its new position. In either case, if you make a point to mark its position carefully and note the time, you will have a rough estimate of its direction and rate of travel. For this sketch, I returned 2.5 hours later to find that the comet had moved about 14 arc minutes to the east-northeast. When you check the new position, carefully compare it to the star framework, and then mark it with a precise "X," "+," or other mark that you will recognize. (**Figure 2.1.17**) Do not forget to note the time. With this step complete, you have a full record of time well spent with the comet. (**Figure 2.1.18**)

Sketching Comets

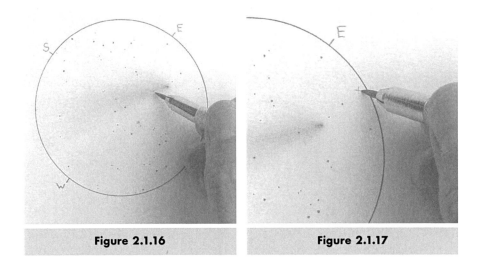

Figure 2.1.16

Figure 2.1.17

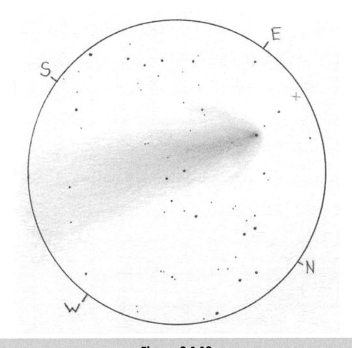

Figure 2.1.18

Astronomical Sketching: A Step-by-Step Introduction

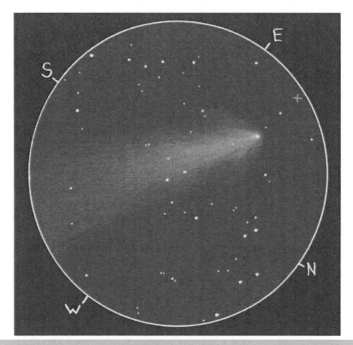

Figure 2.1.19

A positive white-on-black version of the sketch can be seen in Figure 2.1.19.

2.2 Creating a Wide Field Comet Sketch

Every so often, a notably bright comet will grace the night sky, offering an exciting opportunity for detailed binocular observations. Some may even stand up to naked-eye scrutiny. Making an effort to sketch one of these remarkable comets will provide you with a fine record of an event that you will remember for years to come. For a naked-eye or binocular sketch, you may want to print or copy a sky chart of the area so that you can concentrate on the comet rather than the daunting multitude of stars you are bound to see.

This tutorial will feature a binocular sketch of comet C/2004 Q2 Machholz. At the time of the observation, the comet was observable to the naked eye at 4th magnitude. By using 10 × 50 binoculars, I was able to see a thin ion tail and a wide dust tail. This tutorial will demonstrate the use of a preprinted star chart that will serve as a stellar framework for the sketch.

Step 1: Framing and preparing the sketch area If you are preparing your preprinted star background using planetarium software, there are a couple things to consider. First, if you are not sure how large the comet will

Sketching Comets

Figure 2.2.1

appear or where the tail will be pointed, you may want to print a few sheets of the region at different scales and positions and bring them all with you. This will allow you to use the one that is most appropriate for the observation. Another thing to consider is how large your software plots brighter stars. If a test print reveals bright stars that are exceedingly large and you find that distracting, see if your software allows you to change the scaling of star weights. If not, you may want to lay a second sheet of paper over the print and redraw the stars to a more reasonable size. If you find it difficult to see through well enough to trace, try holding the sheets together with a small piece of tape and placing them against a sunlit window, a bright computer screen, or a television. (**Figure 2.2.1**) You can use the same technique with a printed star atlas that you are photocopying or tracing by hand.

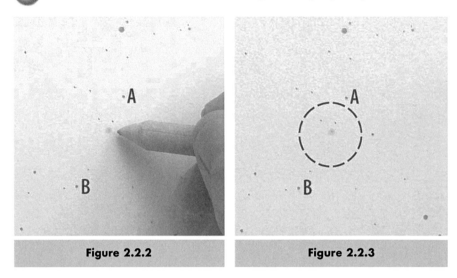

Figure 2.2.2   Figure 2.2.3

Step 2: Marking the comet's position Carefully compare the position of the comet's coma to the surrounding star field. Using a blending stump that is lightly loaded with graphite, mark the position of the center of the coma on your preprinted chart. Jot down the time you plotted this point in your notes. For this sketch, I noted that the center of the comet's coma appeared a little more than a third of the way along an imaginary line from a star marked A to another marked B in **Figure 2.2.2**.

Step 3: Determining the disposition of the coma With this position marked, take time to determine the full extent of the coma. Also note what its brightness profile looks like. Is it very diffuse or is it highly condensed at the center? If it possesses a distinct central condensation or pseudonucleus, is this feature centered or off-centered compared to the entire coma? Use the stars that are visible in the sky and plotted on your chart as a scale to estimate the coma's diameter. For example, in this sample sketch, the coma's diameter appeared to be about two-thirds the distance between the stars marked A and B in **Figure 2.2.3**. The dashed circle represents how I visualized this diameter on the sketch.

Step 4: Sketching the first layer of the coma With the size and profile of the coma in mind, lightly load your blending stump with graphite (see *Using a Blending Stump* on page 153). Take the blending stump and begin swirling it lightly, starting at the point you marked in Step 2. Using a delicate circular motion, proceed outward from the center to define the coma's visible diameter. (**Figure 2.2.4**) As you approach the outer edge of the coma, decrease the pressure on the blending stump to the lightest possible touch. You will usually want these outer edges to fade softly to nothing.

Sketching Comets

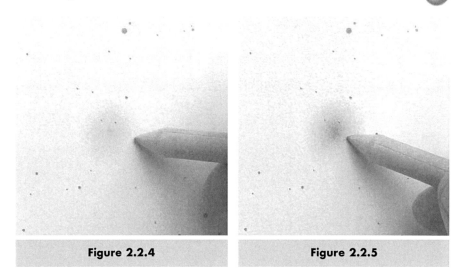

Figure 2.2.4   Figure 2.2.5

Step 5: Sketching the brighter portions of the coma Now that you have applied a base layer that defines the maximum visible extent of the coma, you can fill in any brighter portions you see. Although this brightening will typically manifest itself in the vicinity of the nucleus, you may observe luminous patches in other areas, so keep yourself open to this possibility. Load your blending stump with more graphite if necessary and softly define these brighter portions. (**Figure** 2.2.5) Add them gradually and in layers. Take care not to bring the edges of these layers up to the edge of the base layer if you do not want to give the coma a hard-edged appearance.

Step 6: Analyzing the tail or tails Any visible tails on the comet may be much harder to discern. So take your time, using averted vision, to search for any evidence of them. If you finally get a glimpse of one, pay attention first to which direction it appears to point. Compare this to the stars on your sketch and make a mental note of this. Next, take time to see how far away from the coma you can see the tail stretching. Consider also how wide the tail appears both in the vicinity of the coma and at its trailing end, and whether you can make out any variations in brightness.

Step 7: Sketching the tails In the case of 2004/Q2 Machholz, two tails presented themselves. The first to reveal itself was the slender ion tail that extended like a thin ray from the coma. To sketch this delicate feature, load your blending stump lightly with graphite. Begin by laying down a base layer for the tail. For a long, slender tail, use light, linear strokes to drag the shading for its

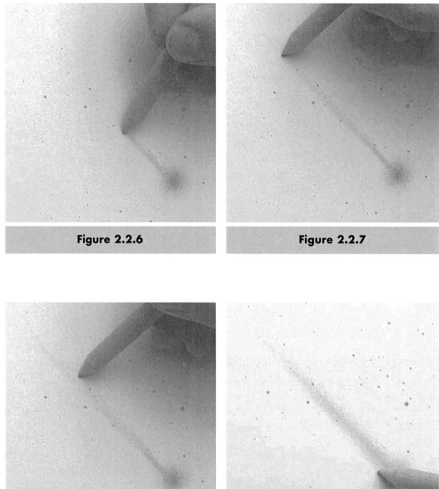

Figure 2.2.6

Figure 2.2.7

Figure 2.2.8

Figure 2.2.9

core away from the coma. (**Figure 2.2.6**) As this core grows fainter, lighten the pressure on your blending stump. (**Figure 2.2.7**)

With this core in place, move back in and add light shading to show the faint outer reaches the tail. (**Figure 2.2.8**) Refine the tail until it matches the relative variations in brightness that you see in the comet itself. (**Figure 2.2.9**)

The second tail appeared as a short, broad fan that was centered about 120 degrees counterclockwise from the ion tail. Begin by sketching from the brighter central region and work outward. Lightly load your blending stump again if

Sketching Comets

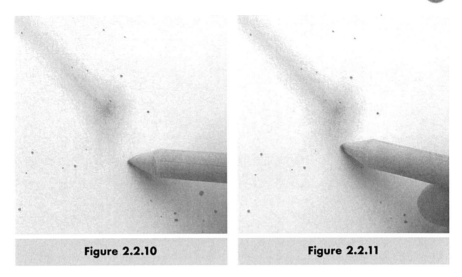

Figure 2.2.10   Figure 2.2.11

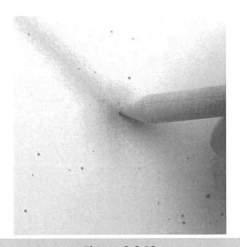

Figure 2.2.12

necessary. Using delicate circular strokes, start a layer of shading beginning with the brighter portion of the tail. Ease up on the pressure so that the shading fades out softly along the outer reaches. (**Figure 2.2.10**) Continue to refine it by reloading the blending stump and darkening the brighter portions. (**Figure 2.2.11**)

Step 8: Finishing the sketch Reexamine the comet and brighten or softly erase any areas that need to be refined to match what you see. (**Figure 2.2.12**) If you drew the stars on your sketch by hand, you may need to redraw any

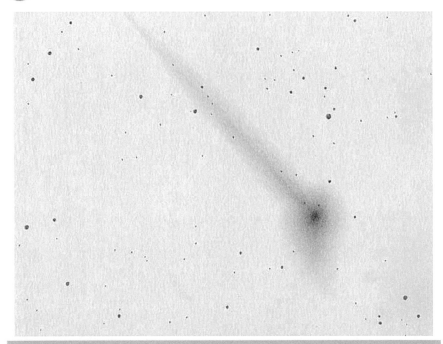

Figure 2.2.13

Figure 2.2.14

Sketching Comets

that were blurred when you shaded the comet. Complete any notes about your observation; the sketch is finished. (**Figure 2.2.13**) An inverted version of the sketch can be seen in **Figure 2.2.14**.

Tips and Techniques

2.3 Assessing Cardinal Directions

One way to check for cardinal directions is to place a conspicuous star at the center of your field of view and turn off the clock drive if you are using one. Then observe which direction this star drifts. The point where it exits the view is west. (**Figure 2.3.1**) If you have an equatorially mounted scope that is polar-aligned, you can nudge the telescope back and forth in declination to indicate north-south. A quick glance at the subtle motion of the top of the scope will tell you whether you are moving the scope toward the north or south celestial pole. If the scope is moving north, stars will enter the view from the north and exit from the south. The opposite will be true when the scope is moving south. I like to use this method since it is faster than watching for drift, especially in low-power views. Once one direction has been indicated, you can automatically fill in the others if you know how your telescope handles an image. Telescopes with an even number of mirrors or no mirrors at all, such as Newtonians and refractors without mirror diagonals, will present "right-reading" views. A right-reading view may be rotated in any direction, but is otherwise normal. Telescopes with an odd number of mirrors, such as Schmidt Cassegrain Telescopes (SCTs) or refractors with mirror diagonals, will present a mirror image. For a right-reading view, working clockwise, the cardinal directions will proceed as follows: north, west,

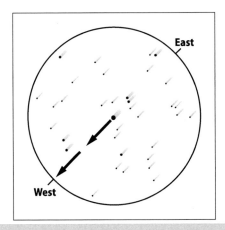

Figure 2.3.1 Star motion proceeds east to west

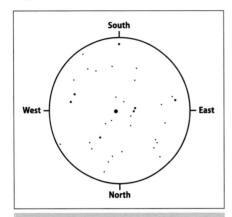
Figure 2.3.2 Cardinal directions in a right-reading view

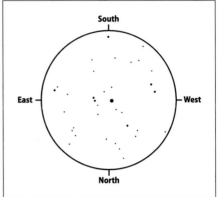
Figure 2.3.3 Cardinal directions in a mirrored view

south, and east. (**Figure 2.3.2**) For a mirror-imaged view, working clockwise, the cardinal directions will proceed as follows: north, east, south, and west. (**Figure 2.3.3**)

2.4 Sketch and Observation Log Sheets

There are a wide variety of methods for recording your sketches and observing notes. No single observing record system will fit the style of every amateur astronomer, so the way you choose to do this will be unique to you. Some observers like to record their sketches and observing notes on preprinted observing record sheets. Others prefer the freedom of sketching and jotting notes on blank sheets of paper, using sketch circles or none at all, as the situation demands. Others combine these two methods in various ways. If you are new to sketching, try a few methods and find what suits you best. On page 181, you will find a condensed deep sky observing form with space for a sketch. A variety of other observing forms can also be found online.

If you want to add a sketching circle to a blank sheet of paper, there are a few things you can do. You can find a cup, can, or bottle with a suitable diameter and trace around this to create a circle. A compass is another obvious and handy instrument for making sketching circles. However, the most useful tool I have found so far is a plastic circle template with circle diameters ranging from 1.25" to 3.5" (32 mm to 89 mm). (**Figure 2.4.1**) Such a template will allow you to quickly trace circles without using the awkward arm positions necessary to sketch around a cup, can, or bottle. It also avoids the pinprick that some compasses can leave at the center of their circles.

Whichever method you use, do try to capture at least the following information: object observed, date and time, instrument used, eyepiece and filter used, magnification, and atmospheric conditions. Not only will this information assist

Sketching Comets

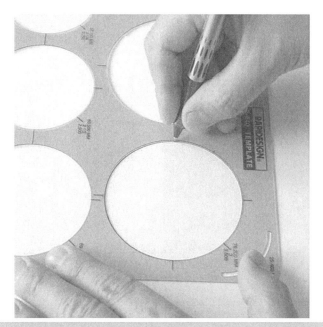

Figure 2.4.1

you in recalling the details of your observing sessions, it will also help you prepare for future observations as you consider what combinations of equipment and conditions led up to the view you were able to sketch. This information is also valuable to other amateur astronomers if you decide to share and compare your observations.

There are a wide range of papers to choose from for your sketches and observation records. A few important things to consider are the weight, texture, and acid content of the paper. Heavier papers are more likely to handle better and be less likely to wrinkle under humid conditions. Papers are also offered in a wide range of textures, from smooth to rough. Smooth papers are well suited for detailed work, but may not always accept multiple layers of shading because the "tooth" of the paper fills in more quickly. Rougher papers tend to accept more layers of shading, but will also display heavier texture to any shading you add. Finally, if possible, try to use an acid-free paper to increase the longevity of your sketches. Papers with acid content will yellow and deteriorate over time. There are no definitive answers on what type of paper you should use, and you should experiment to see which you prefer.

Finally, you will want to consider how you want to store and protect your finished sketches. Some astronomical sketchers keep their work indexed and stored in vertical file folders, flat files, or plastic sleeves. Others keep their sketches in 3-ring binders or in pre-bound sketch books. The archival possibilities are numerous. Whatever method you use, you may wish to apply a spray fixative to your sketches to help protect them from smudging and scuffing. If your sketches

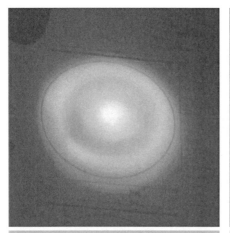

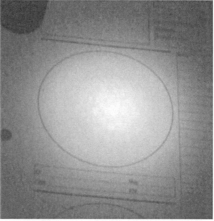

Figure 2.5.1 Distracting flashlight patterns

Figure 2.5.2 The same flashlight beam with a piece of wax paper behind the lens

are stored securely and seldom disturbed, a fixative may not be as crucial. However, if you regularly browse, rearrange, or show your work, applying fixative can be important insurance for your hard-earned memories. The need for a fixative becomes even more critical if you use charcoal or pastels in your sketches, since these are much more susceptible to wear and tear than graphite.

Spray fixative can be purchased from most local or online art supply stores. When applying the fixative, be sure to do so in a well-ventilated area—outdoors if possible—and to spray one to two feet away from your sketch, using smooth even strokes. It is better to apply two or three light coats of fixative than to apply a single soaking coat. Allow the fixative time to dry between coats, and before touching the sketch area or sandwiching it between other sheets of paper. With a little planning and care, your astronomical sketches will be preserved for you to enjoy for a long time to come.

2.5 Sketching Faint Objects in Low Light

When you decide to sketch faint, delicate objects that hover on the threshold of vision, you will face a dilemma. To see these objects well, your eyes need to be adjusted to the dark. However, in order to sketch these elusive cosmic quarries, you need some form of light to see what you are doing. How can you reconcile these two competing needs?

The first thing to do is invest in a red flashlight with a dimmer switch, available from most amateur astronomy suppliers. Red lights can sometimes project irregularly lit patterns onto your sketch. (**Figure 2.5.1**) This can be very distracting when you are trying to sketch a faint object. One way to help diffuse this pool of light (**Figure 2.5.2**) is to place a strip of masking tape over the lens of the flashlight. I have found that opening the flashlight up and placing a piece of wax

Sketching Comets

paper behind the lens also works very well. Having a dimmer switch on your red flashlight is very important. When sketching, you must endeavor to keep the light turned down as much as possible. As your eyes adjust to the dark, that faint light will gradually become useful enough to accurately render your sketch—at least when it comes to stars.

The situation becomes a little trickier when rendering faint nebulosity. The faintest settings on your sketch light can make it very difficult to discern the feeblest smudges of blended graphite on paper. With the light at such a low level, you may end up overdoing the darkness of the shading simply because you cannot see it on the sketch. This can be especially true when you are taking your first steps in the realm of astronomical sketching. This struggle will become more manageable with practice sketching in the field.

Turn the light up a little as you work toward this comfort level. As you get better acclimated to your sketching tools and techniques, the brightness of the light can be dropped back down again. No matter how dim you keep your light, it is still an artificial light source and will affect your dark adaptation to one degree or another.

To compensate for this, the most important technique is to keep your eye glued to the dark recesses of your eyepiece as long as it takes to maximize your dark adaptation. For particularly elusive subjects, you may want to turn your sketch light off completely while you are absorbed in the view. Pull a hood or towel over your head and eyepiece to block stray light. Concentrate carefully and at length on the view in the eyepiece, burning the image into your mind as strongly as possible. Once you have the image firmly in mind, drop down to the sketch, turn the light back on, and add the shading quickly while the image is still in your mind's eye. Try to spend as little time as possible with your eyes exposed to the red light. Once you have applied as much as you can remember, return to the eyepiece, letting your eyes adapt again before going back to the sketch and starting the process over again.

This may seem frustrating if you are uncomfortable putting pencil to paper to begin with. You may feel the need to spend as much time as possible at the sketch, with the light turned up brightly, working your way through the mechanics of the whole process. And you know what? That is perfectly fine. Work under the conditions you need until you are comfortable with the process. However, keep these points in mind as you gain confidence. Try to apply them with every sketch and you will find that it becomes easier with time.

Figure 2.5.3

One final issue is how to hold your red flashlight. If you are able to sit while observing, you can keep the clipboard in your lap and hold the flashlight in one hand while you sketch with the other. (**Figure 2.5.3**) However, if you must stand to observe, the solution requires a bit more creativity since you must now hold the clipboard with your free hand. A simple solution is to lay the flashlight on its side on the clipboard and aim it at your sketch. (**Figure 2.5.4**) This can be a frustrating solution though, because the ray of light will fade from light to dark across the length of your sketch. The flashlight will also tend to shift position if you do not keep the clipboard still and level. Another option, if the flashlight has a lanyard, is to hang it from a convenient structure such as your telescope's eyepiece. (**Figure 2.5.5**) You can also try rubber banding it to the head or arm of a camera tripod and aiming where you need it. (**Figure 2.5.6**)

One solution that I found particularly useful was to purchase a spring clamp, a hose clamp, and a gooseneck lamp with a clip. I removed the lamp head and wiring from the gooseneck lamp. I then used the hose clamp to attach the spring clamp to the loose end of the gooseneck. (**Figure 2.5.7**) The red flashlight is then held in the

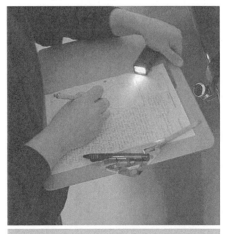

Figure 2.5.4

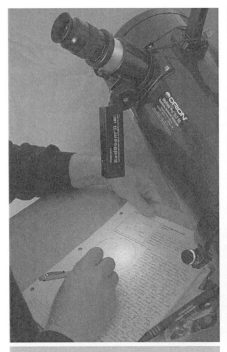

Figure 2.5.5

Figure 2.5.6

Sketching Comets

Figure 2.5.7

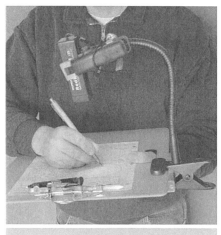

Figure 2.5.8

spring clamp, and the gooseneck is clipped to the clipboard. (**Figure 2.5.8**) You may need to slip a narrow item such as an empty cassette tape holder beneath the clipboard so that the gooseneck clip has a thicker surface on which to attach itself.

Another lighting solution that some amateur astronomers find useful and convenient is a red LED headlamp. A headlamp can keep your hands free, and point the light right where you need it. However, keep in mind that even a headlamp with adjustable brightness settings may still be too bright at its dimmest setting. If you find this to be the case, judicious application of masking tape, or electrical tape with small holes in it may help dim the light down to a manageable level. Other lighting solutions can also be found online in most amateur astronomy forums.

CHAPTER THREE

Sketching the Sun

Star Light, Star Bright

To Earth and its inhabitants, our Sun is the key that sustains life. To our solar system, the Sun is the key to its creation. Exploring our parent star opens an opportunity for us to witness phenomenal events, all associated with the production and transportation of energy. With minimal equipment, not only can we observe these fascinating events, we can record them, follow trends, and capture them through sketching! And if that is not incentive enough, we enjoy the warmth of daylight while doing it.

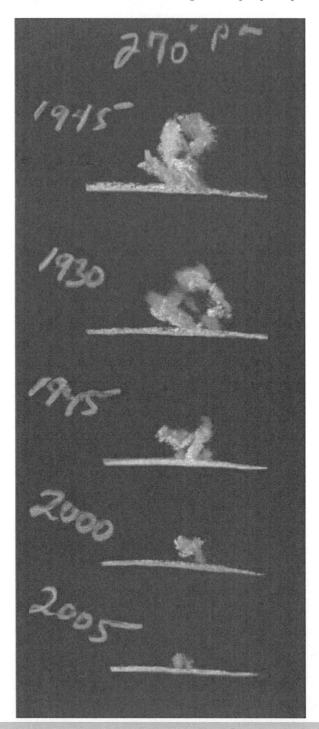

Figure 3.0.1

Sketching the Sun

Figure 3.0.1 shows a rendering sequence of a "Disparition Brusque" captured while observing with an H-alpha filter system on April 22, 2006 over the course of 50 minutes.

Figure 3.0.2

Solar Observing Tip

Solar features are sometimes hard to spot without adequate observation time. It is similar to attempting to locate an airplane in the vast sky. Pilots are trained to scan the sky in sections for other aircraft. I use this same approach while observing the Sun. At the beginning of the observation, I start at the lower section of the solar disk, working my way east to west, gradually moving upward until I have had a very good look at the entire "surface." If viewing in H-alpha with my Personal Solar Telescope, I constantly tweak the focuser and tuning ring while scanning, and then make my way around the entire outer limb. The amount of detail that can be found is sometimes quite amazing. (**Figure 3.0.2**)

Warning!

Although it is assumed that you are familiar with the dangers of solar observing, remember that simply reducing the amount of light visible from the Sun is not enough to ensure protection. We must proceed with caution and use the proper filters suitable for our telescopes and binoculars and/or the proper techniques, as described in the following tutorials.

Photosphere

The Sun is basically made up of six layers, an ever-changing ball of energy starting with nuclear fusion at its core. The energy created by reactions in its core then travels outward through the radiation zone, thus entering the convection zone. Here, the bubbles of condensed hot gas expand and rise toward the fourth layer, the "surface" of the Sun called the photosphere. It is at this layer that we commence our tutorials.

There are many ways of rendering an accurate sketch, and the tutorials in this chapter merely reflect a few examples to get you started. Once you begin, you may wish to explore other methods that will suit your style somewhat better. For white light or projection sketching, the size of the circle is a matter of preference. I find that sketching larger assists in creating detailed work and recommend at least a 2″ to 5″ diameter circle.

3.1 Basic White Light Sketching

Perhaps one of the easiest renderings of the solar disk is done at the eyepiece with a white light filter. There are many types of these filters available, such as film, glass, and metal. Whichever you decide on, take care to purchase a filter that meets the manufacturer's recommended specifications and only use the kind that connects firmly onto the opening of the optical tube assembly (OTA). The downfalls of sketching directly at the eyepiece are inadvertently sketching features larger than they actually are and/or slight misplacement of them. It is helpful to imagine a grid line on both the solar disk and on the sketched disk while applying features. You can always enlarge the sketched features, but it may not be easy to erase them cleanly if you want to reduce their size.

- Finder scopes should have the objective ends covered so that you will not accidentally look through it during your session. Also, they sometimes have crosshair eyepieces that can be melted or burned by the focused solar image.
- You may need to stop down the aperture of your telescope to 60–100 mm (or under 3″–4″), depending on the type of filter being used.
- Do not look up at the Sun to align the telescope. Instead, watch your telescope's shadow. Once the shadow becomes minimized and circular, the scope will be fairly lined up.

This simplified sketch can be quickly rendered for a daily sunspot recording.

Suggested art supplies

- Sharpened #2 pencil
- Compass
- Eraser
- Paper

Step 1 Begin by creating a circle with a compass.

Step 2 If penumbrae exist, add them gently with a blending stump that has been rubbed into charcoal. This may also be done lightly with a pencil if you prefer and then blended. If needed, gently add a detailed border to the penumbra with a sharp pencil or charcoal stick. (**Figure 3.1.1**)

Step 3 The umbrae may now be added in the sunspot groups over the penumbrae using a pencil/charcoal stick. You may wish to increase the magnification at this stage to observe more detail. Start by adding the larger ones first. Pay close attention for irregular shapes and placements. Sometimes you may see a displaced umbra along the limb, which is a prime example of an irregular shape. (**Figure 3.1.2**)

Step 4 Once the foundation is rendered accurately, add the smaller spots. (**Figure 3.1.3**)

Sketching the Sun

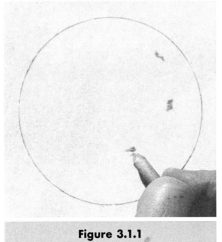

Figure 3.1.1

Figure 3.1.2

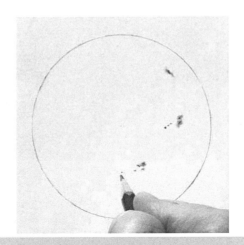

Figure 3.1.3

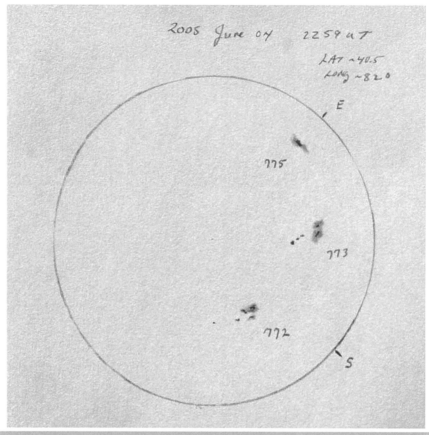

Figure 3.1.4

Step 5 Add notations relevant to your observation. (Figure 3.1.4)

3.2 Projection Sketching

Benefits of projection observing include access to viewing for a group of people rather than just one at a time. Also, have you ever noticed how difficult it can be to place the sunspots accurately on your paper or not to get carried away by making the sunspots too large? Projection sketching allows a more accurate recording.

Although projection sketching is an alternative to safe direct-observing techniques, there are hazards. As solar observers, we are aware of the cautions; nevertheless, it is to our benefit to review them before getting started. As well as the cautions noted in the White Light Filter Technique section, an additional precautionary list is provided below:

- Some telescopes are not designed for projection use, and if damaged, they might not be covered under the manufacturer's warranty. Therefore, it is suggested that you check the suitability of this type of observing with the manufacturer first.

Sketching the Sun

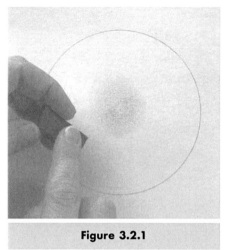

Figure 3.2.1

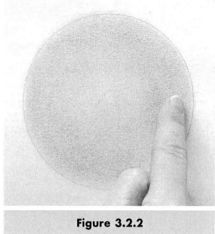

Figure 3.2.2

- Stop down the aperture of your telescope to 60–100 mm (or under 3″–4″).
- Use simple, inexpensive, low-powered eyepieces. Even though the telescope is stopped down, there will still be a lot of energy condensed at the eyepiece that could overheat and damage it. The more expensive eyepieces tend to have more lenses, which increases vulnerability to damage. Also, keep in mind the hazards of using crosshair eyepieces.
- **DO NOT LOOK THROUGH THE EYEPIECE TO FIND THE SUN.**

Suggested art supplies

- Sharpened #2 pencil
- Compass
- Eraser
- Paper (I have used Rite in The Rain Paper for this tutorial. More information on this paper can be found in Chapter 1 in the Charcoal Sketching Tutorial.)
- Stick of black compressed charcoal, preferably dark tone
- Two sizes of soft blending stumps
- Sharp utility knife to sharpen the eraser to a fine point
- Sandpaper to sharpen the blending stumps
- Clipboard with brackets or an extra tripod to support the paper firmly against the telescope

Step 1 I tend to begin the observation using a white light filter and observe the solar disk through the eyepiece for any unique observations such as limb darkening. Using a compass, create a light circle on white paper. In this example, a background was developed within that circle by rubbing my fingertip onto a stick of charcoal and applying the charcoal onto the paper with my finger. (**Figure 3.2.1**) If the paper is textured, granulation is achieved. Start with the innermost section of the disk and work your way outward, allowing the disk to become darker toward the limb areas. You may need to apply charcoal to your fingertips several times during this process. (**Figure 3.2.2**) To ensure a clean, crisp

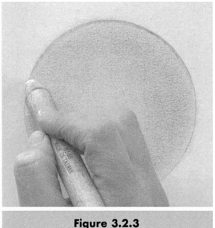 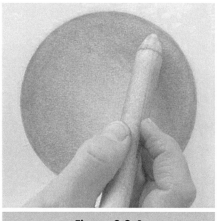

Figure 3.2.3   **Figure 3.2.4**

limb line, rub a blending stump into the charcoal to apply the limb darkening. (**Figure 3.2.3**)

Step 2 Another layer can be added using the technique in Step 1 if a darker background is desired. I have used a thick blending stump in **Figure 3.2.4** to soften the extra layer.

Step 3 Take the white light filter off of your telescope and position your paper behind the eyepiece holder and adjust the depth so that the solar disk fills the circle you have created, making sure that the paper is perpendicular to the optical axis. If it is not perpendicular, you will see an oblong solar disk. It helps to mount the paper onto a clipboard, either braced by a tripod or by mounting brackets so that the disk is a constant size while you sketch. It may help contrast if you create shade by fixing a piece of cardboard around the shaft of the eyepiece holder, keeping unwanted light off of the projection on your paper.

Step 4 Using a very sharp pencil, trace the outline of the sunspots, starting with the umbra first and then adding the fainter penumbras lightly if visible. (**Figure 3.2.5**)

Step 5 The sunspots can be filled in away from the projection view in a more comfortable position by adding a white light filter to the OTA and finishing your sketch at the eyepiece.

Step 6 To finish the sketch using a white light filter, you will notice that the existing sketch is mirrored and flipped upside down. Keep that in mind while filling in the detailed features and trust what you have already added to the sketch during projection. I flip the sketch top to bottom to assist with that aspect of the orientation. Using a sharpened eraser (**Figure 3.2.6**), rub out the existing background to create faculae. Try not to completely erase the existing sunspots during this process. (**Figure 3.2.7**)

Step 7 Sharpen a slender blending stump as indicated in **Figure 3.2.8** and then rub it into the stick of charcoal (**Figure 3.2.9**). Apply the fainter features such as the inside of the penumbra or perhaps more limb darkening. (**Figure 3.2.10**)

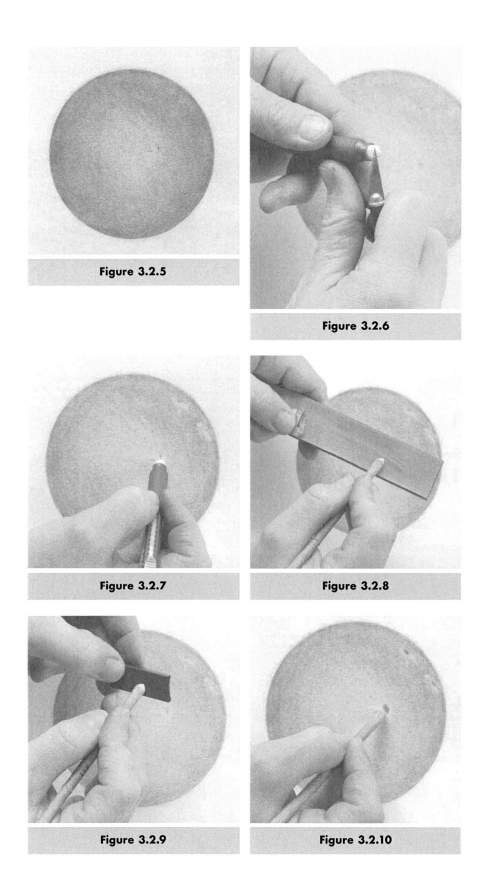

Figure 3.2.5

Figure 3.2.6

Figure 3.2.7

Figure 3.2.8

Figure 3.2.9

Figure 3.2.10

Figure 3.2.11

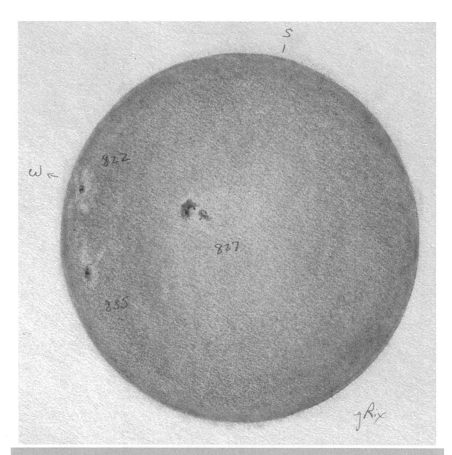

Figure 3.2.12

Sketching the Sun

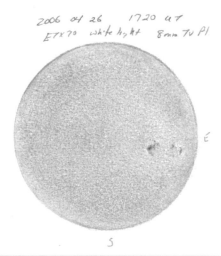

Figure 3.2.13

Step 8 With a sharpened pencil, define the shapes of the sunspots if needed for accuracy (making sure that you do not enlarge them) and add any smaller sunspots within the group that may not have been detectable through projection. It may also be beneficial to increase magnification if conditions permit for more detail. (**Figure 3.2.11**)

Step 9 Add any notations concerning your observation to the sketch. (**Figure 3.2.12**)

Note that if you wish to create an in-depth white light filter sketch with the background solely through the eyepiece, omit Steps 3–5 of the projection tutorial. Since sunspots can be added later with an eyepiece sketch, I would suggest creating the background first, erasing out the faculae areas, then add penumbra, with the umbra added last. (**Figure 3.2.13**)

Chromosphere

Chromos is the Greek word for color. This fifth layer of the Sun is approximately a 2000-km-thick layer of gas that can easily be viewed with narrow-band filters. An example of this special filter is the H-alpha filter that blocks out all other wavelengths of light, only letting the red light through. As with all solar features, the temperatures within the Sun will permit the contrast for distinguishing these features. The cooler filament features will appear darker and the hotter features, such as plage, will appear much lighter. Narrow-band filter sketches can be quite a rewarding next step.

Astronomical Sketching: A Step-by-Step Introduction

> **Solar Observing Tip**
>
> There is a little tidbit of advice I would like to add at this time. Consider the drastic change in lighting that your eyes adjust for when you walk in a dark room after standing outside on a very sunny day. You get that same effect at the eyepiece when you back away from the darker view of H-alpha to look at your sketch during a bright day. It may take a few minutes for your "night vision" to adjust back to the darker view through the eyepiece. I rarely notice this adjustment when sketching using white light filters because, in both cases, the views are very light. So be prepared to wait a few minutes at the eyepiece while your vision adjusts to bring clarity to the features you are observing with narrow-band filters. A dark cloth to cover your head and eyes with or a sun shield will dramatically improve the details you could see while observing.

3.3 Ha Filter Sketching: Prominences

Here is where H-alpha sketches come into their glory. By using black paper, you eliminate the unevenness of creating a black background around the limb by hand, as well as saving time. The results have been the truest to the views I have been able to produce through an H-alpha filtered telescope. This is perhaps my favorite medium for recording solar prominences, with the black charcoal on white paper as a close second. In either case, they both require the same sketching technique, with inverting the charcoal sketch in the final step as the only difference between the two.

Suggested art supplies

- Compass with white pencil
- Eraser
- Black sketching paper (I recommend 60-lb. acid-free Strathmore Artagain paper.)
- White chalk (A single Conté Crayon was used in this tutorial; however, Blanc B and Blanc 2B will give you a nice blend of tonality.)
- Slender blending stump
- Sandpaper to clean the blending stump

Step 1 A shallow arc is created with a compass inserted with a white pencil. Sometimes I have more success with simply drawing the arc freehand using the white Conté. The size of the arc is determined by how large of a prominence you wish to create. It may be helpful to create the arc matching the orientation of the prominence on the disk, but I generally draw the arc horizontally on the paper. Add extra chalk below the arc to render a portion of the limb if you prefer.

Step 2 Sharpen the edge of the arc with a blending stump and then continue to blend gently, starting with the darkest area of the limb and gradually getting lighter toward the center of the imaginary solar disk. (**Figure 3.3.1**)

Step 3 Draw the brightest areas of the prominence directly with your chalk. This will be the foundation of your prominence for accurately rendering the rest of your sketch. (**Figures 3.3.2 and 3.3.3**)

Sketching the Sun

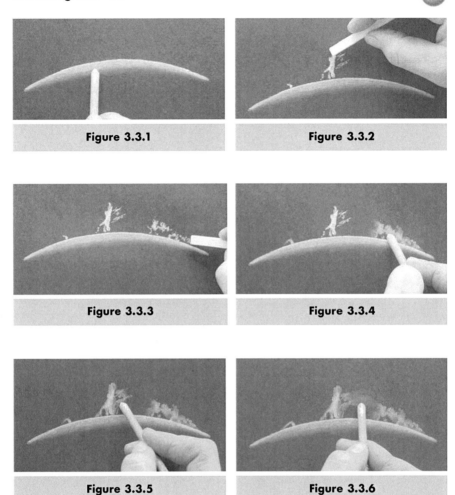

Figure 3.3.1

Figure 3.3.2

Figure 3.3.3

Figure 3.3.4

Figure 3.3.5

Figure 3.3.6

Step 4 Gently blend those areas with a clean stump, spreading them outward as wisps of etherealness. By working in small, quick strokes while blending, you will have more contrast control within the brightest areas. Refer back to the prominence through the eyepiece as needed during this process and be careful that you do not overextend your blended markings. It is easy to make these areas larger on your paper than what they truly are. Study the prominence through the eyepiece. You will soon see definition within the brightest areas, faint branches reaching to their neighbors, and translucency as a backdrop. Take advantage of blending to capture these qualities. (**Figures 3.3.4** and **3.3.5**)

Step 5 Excess chalk on the blending stump can be used to render the translucent areas. Mark a spot in a test area outside the sketch section first, making sure that the translucent area will not be rendered too dark. You can rub off any excess chalk on the test area. If you need more chalk, steal some from the limb area or rub your blending stump against the chalk to reload it. (**Figure 3.3.6**)

Figure 3.3.7

Figure 3.3.8

Figure 3.3.9

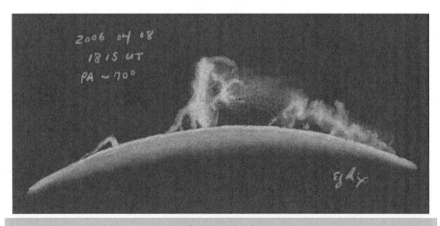
Figure 3.3.10

Step 6 Repeat Steps 3 and 4, adding an additional contrast layer over the foundation. This time, be gentle with the blending as this step involves bringing more contrast to the prominences. (**Figures 3.3.7** and **3.3.8**)

Step 7 A little bit of chalk goes a long way on smoother paper such as the 60 lb. or less. If you struggle with using the blending stump on heavier chalked areas,

Sketching the Sun

use your finger to lightly tap the areas you want softened. By tapping the area, you smooth out the rougher areas without ruining the contrast of the second layer of chalk. (**Figure 3.3.9**)

Step 8 Mark the approximate position angle, date, time, and any other notes you wish to include on your sketch. (**Figure 3.3.10**)

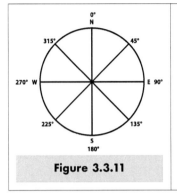

Figure 3.3.11

Prominence Position Angle Tip

Position angles of prominences can be recorded several different ways. I prefer to use the diagram in **Figure 3.3.11** to record the position of the prominences on the solar limb.

Prominences: Colored Chalk on Black Paper

Using a magenta colored Conté crayon (or similar colored chalk) on black Strathmore paper, the same technique can be applied for a colored prom sketch as in **Figure 3.3.12**.

Figure 3.3.12

Prominences: Black Charcoal on White Paper

As with black-on-white Deep Sky Object sketching, you can create a similar sketch with prominences using charcoal on white paper. Whether you will turn the sketch into a negative image after scanning is a matter of personal preference. However, it does bring out the striking contrast and airiness of close up prominence sketches. In either case, you are sketching a dark feature on a light background, where in reality, what you are viewing through the eyepiece is a lighter area on a dark background. Thus, we are working in negatives. Again, using a black cloth to block out excess light will assist you tremendously in seeing the delicate flows of these features. Follow the above tutorial using charcoal on white paper instead of white chalk on black paper. **Figure 3.3.13** shows what this type of sketch looks like when inverted.

An example of improved observing and sketching skills can be seen by comparing the prominence sketches in this tutorial with my first colored close up prominence sketch **(Figure 3.3.14)** using a Personal Solar Telescope (PST) with a 12-mm eyepiece. I am still using a PST today with most of my close up prominence sketches rendered using either an 8-mm TeleVue Plössl or one of my 12-mm eyepieces.

Figure 3.3.13

Figure 3.3.14

Sketching the Sun

Figure 3.3.15

Here is a more recent close up prominence sketch using the PST with an 8-mm TeleVue Plössl (**Figure 3.3.15**). Steady viewing and patience at the eyepiece will allow the details to appear and a dark cloth is always used over my head and eyepiece to block out excess light. I spend more time viewing a single prominence than it takes me to actually sketch several of them.

3.4 Ha Full Disk Sketching

Surface details are the key features in full disk renderings. The prominences are diminutive compared to the large solar disk; however, they are still added for a complete observation record.

Suggested art supplies

- Compass with red or magenta colored pencil
- Eraser (I prefer a white vinyl pencil type eraser that can be sharpened easily with a utility knife.)
- Black sketching paper (60-lb. acid-free Strathmore Artagain paper was used in this tutorial.)
- Colored chalk (Three shades of Conté Crayons represented the shades of the solar disk for this tutorial.)
- Variety of blending stumps
- Sandpaper to clean and sharpen the blending stumps
- Deep red colored pencil
- Terry cloth rag or paper towel

Step 1 Draw a circle on your paper using a compass equipped with a dark red pencil. Generally, I create a circle 3″–5″ in diameter. Using a dark red/magenta

Astronomical Sketching: A Step-by-Step Introduction

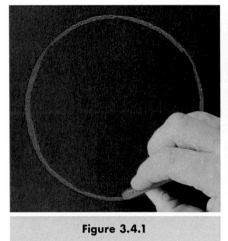

Figure 3.4.1

Figure 3.4.2

Figure 3.4.3

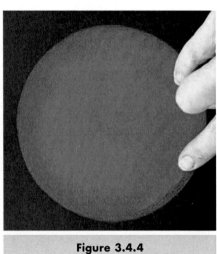

Figure 3.4.4

chalk, trace the inside of the circle. Try to match the color to the limb coloration that you see through the eyepiece. (**Figure 3.4.1**)

Step 2 The surface color usually has a magenta hue to it; however I have noticed that it never really appears to be the same color twice. Sometimes a brighter orange or brilliant red color is apparent. I chose three colors from my Conté collection to represent the colors I saw for the sketch in this tutorial. Beginning with the color in the center of the disk and working your way to the limb, apply the chalk, layering over each other with each color change. (**Figure 3.4.2**)

Sketching the Sun

Figure 3.4.5

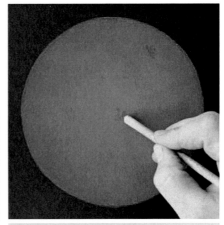

Figure 3.4.6

Step 3 Using your finger or thick blending stump, blend the colors in a circular motion starting in the center of the disk. Work your way to the limb, pushing any excess chalk dust to the bare areas until the whole disk is colored. Add another layer of chalk if the color is not quite right. In this example, I have added a touch more magenta, as the center of the disk was more uniform with the outer colors. (**Figure 3.4.3**)

Step 4 If you run out of excess chalk dust to complete the disk, add a thin layer of chalk directly to the inside limb. Then blend with your finger lightly, being careful not to go outside the circle. Blow off any excess chalk dust and then wipe around the edge of the disk with a dry paper towel or cloth to remove any smears. An eraser is used for a clean limb edge. (**Figure 3.4.4**)

Step 5 Sometimes you may notice limb darkening. Although you can use colored chalk to render this type of phenomenon, take advantage of the dark paper by lightly rubbing the excess chalk off the limb. Use your finger as if you are blending. (**Figure 3.4.5**)

Step 6 Scan the solar disk slowly, moving side to side, working your way bottom to top. (**Figure 3.0.2**) View one section at a time, allowing your eyes to adjust. If you scan too quickly, you run the risk of missing valuable additions to the sketch. Start with the backdrop of the surface details. You have already added the basic colors of the solar disk. Plage areas are next. Smoother paper does not always allow for additional layers of chalk to be applied with a natural effect. If you were to use white paper instead of black paper, the plage areas could be rendered by simply erasing the chalk in those areas and the white paper beneath would be your rendering of these features. However, because you are using black paper, plage areas are easily created by rubbing off the chalk with a blending stump as in **Figure 3.4.6** and then adding the new color

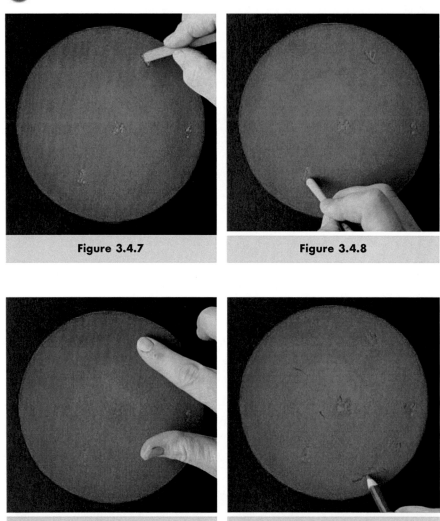

Figure 3.4.7

Figure 3.4.8

Figure 3.4.9

Figure 3.4.10

that represents plage. I have used an orange/yellow Conté crayon for this step. (Figure 3.4.7)

Step 7 Blend the new color very by lightly tapping those areas with a narrow blending stump as shown in **Figure 3.4.8**. Excess rubbing will actually remove the new layer instead. Then lightly tap with your fingertip to soften the blend. (Figure 3.4.9)

Step 8 Take care to render the faintest of filaments very lightly, following the spindly directions they lead. Some of the filaments are dark, thicker, and jagged, sometimes forking off in a new direction with a very faint line that is barely detectable. Each have unique characteristics and need to be rendered accordingly.

Sketching the Sun

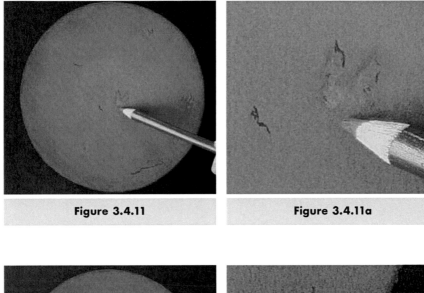

Figure 3.4.11

Figure 3.4.11a

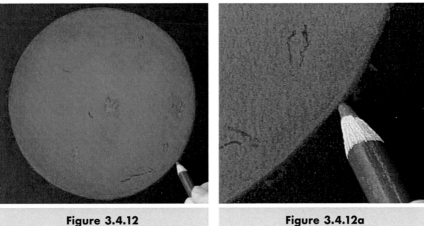

Figure 3.4.12

Figure 3.4.12a

The filaments and sunspots are rendered with a deep red or orange colored pencil. As you use the pencil, note how it has a tendency to rub off the layers of chalk as you do so. It has a nice effect with the black paper underneath slightly showing. (**Figure 3.4.10**)

Step 9 Notice how I darkened the sunspot grouping in the center of the disk. The grouping had a darkened area around the plage. This would have been easily overlooked had it not been for a moment of steady seeing and great transparency. If you notice areas such as this, lightly shade them in with your dark red pencil. (**Figures 3.4.11** and **3.4.11a**)

Step 10 Lightly sketch in the prominences using your red or magenta pencil. Try to be accurate with size and shape. (**Figures 3.4.12** and **3.4.12a**)

Figure 3.4.13

Step 11 Add any notations concerning your observation to the sketch. (**Figure 3.4.13**)

Corona and Solar Phenomena

The corona is the outermost layer of the Sun and can be viewed with solar-eclipse-like circumstances. Other ideas for capturing Sun-related phenomena are auroras, sundogs, arcs, rainbows, or even the ion tails of comets being repelled by solar wind. Think outside the box and experiment. Above all, enjoy the experience.

CHAPTER FOUR

Sketching the Planets

Many people feel that observing planets requires special skills, superexpensive optics, and perfect seeing conditions. Although all of these ingredients are important, having fun and rewarding observations can still be had if you have a basic understanding of how to use what you have. I would like to start with some rudiments before I touch upon the use sketching materials and various techniques.

The sketch tutorials that follow show the cardinal directions of Saturn, Jupiter, and Mars with south at the top and north at the bottom. The right side of these sketches is the preceding side that is rotating away into evening. To the left is the following side in which planetary features emerge from night into daytime. This orientation is typical of a view through a refractor telescope using a star diagonal as well as any telescope that has an odd number of reflecting surfaces. A Schmidt-Cassegrain telescope with a star diagonal has three reflecting surfaces. Newtonian reflectors have an even number of reflecting surfaces, so the image will appear reversed horizontally.

Seeing Conditions Good planetary observing does not require dark transparent skies and an acute acquired night vision. It does require that you have a basic knowledge of the environment of your observing sight. In time, you will get an understanding as to what time or under what conditions the air is at its steadiest. You do not want to observe close to any sources of heat or air turbulence. Air conditioners, houses and their roofs, asphalt, and so forth, are notorious for disturbing air. I believe that most "seeing" occurs in the first 50 feet of air above ground. Additionally, your body heat can ascend and pass in front of your telescope's line of sight and distort your view. If it is cold outside while you observe, you should insolate your body enough to trap the heat that will come off you, especially around the head and hands and when using a Newtonian telescope. For this, a hat and gloved tracking hand is useful in combating any heat waves so that you would be able to see through your telescope's eyepiece. For telescopes that are mounted closer to the ground, like my reflector on a man-made surface, I cover the ground with a heavy-duty painter's tarp to keep air currents from rising off the ground and distorting my viewing.

Magnification Changing eyepieces for a slightly different magnification, changing a filter, or touching up the focus may not actually achieve much in a measurable way regarding observed details. However, when observing one object for an extended period of time, eyepiece or filter changes may be beneficial to you since it can shock your eye out of its complacency.

Magnification is sometimes dependent on seeing conditions. When trying to choose the best magnification, there are general rules regarding selection. One of these relates to your telescope's aperture. If your telescope's optics are decent, then 37× per aperture/inch can get you close to the highest magnification while maintaining good resolution. This means that a 6″ telescope is best at 222× magnification. A 10″ telescope would do well at 370×. This would work well for Mars and Saturn. However, you may find that Jupiter will only do well at somewhere between 200× to 300×.

Another method, which I use, for choosing a proper magnification is to go by the size of what is called the exit pupil. An exit pupil's size is produced by your telescope's focal ratio divided by an eyepiece's focal length in millimeters. It is generally recognized that a telescope and eyepiece combination should produce an exit pupil of 1 mm in diameter for the best magnification and resolution. If you follow this method, it means that a telescope with a focal ratio of 8 (f/8) would operate best with an 8-mm eyepiece. Since everyone has optics, eyes, and seeing conditions that are different, this rule can be adjusted as needed. I can comfortably use exit pupils that go down to 0.4 mm in diameter.

At high magnifications, you may see dark little squiggly shapes floating around inside your eye while observing. These are called floaters. The presence of floaters can affect observers to varying degrees as magnifications climb. Floaters make themselves more abundant if you look down for a while. To combat this, I find that tipping my head back and looking up for a couple of seconds clears a lot of them away.

Other Tips Averted vision is another key observational technique that I use all the time. This is nothing more than occasionally looking a little off to the side of the object you are observing. By doing this, you stimulate a different part of your eye's retina. This will increase your seeing sensitivity. This is a technique that deep sky observers use frequently.

Sketching Tools The tutorials that follow will only show black-and-white or monochrome examples. I believe that this is the best way to start. Good results can be achieved with a minimum of materials. The materials I used in the fol-

Sketching the Planets

Figure 4.0.1

lowing tutorials are quite simple. **Figure 4.0.1** shows my actual setup for sketching at the telescope.

- Graphite pencils: 4H and 2B
- White colored pencil
- Eraser (I use the kind that can be pushed on the end of my pencil. A blade is used to create a sharp chiseled edge.)
- Eraser shield
- Clipboard (I place a smooth piece of cardboard onto the clipboard's surface upon which I place my drawing paper.)
- Light source (A battery-powered musician's sheet music light is great for illuminating the drawing surface. It has a built-on clip so it can be moved to any location on your clipboard. For planets, red light is not necessary.)
- Smooth paper
- Blending stump (I believe that this item is optional since I feel that blending with my fingers offers more control.)

4.1 Sketching Saturn

Our first sketching tutorial shows every step taken to render Saturn. It is best to start with a template that accurately portrays the outlines of Saturn's globe and rings. I have provided one in the Appendices section of this book. This template shows Saturn's inclination at the time of this writing. Additional templates showing different inclinations of Saturn through time are available on the Internet. Although you may find Saturn a bit easier to sketch, observing any of its details may prove more difficult than those presented on Jupiter or Mars. Some of these details may appear quite different from time to time. For example, the shades of tonality appearing in Saturn's Polar Region may appear uniformly toned or sometimes appear as belts. Additionally, the Polar Region may have a dark cap or a semicircular clearing. The Southern Equatorial Belt, Saturn's most prominent disk feature, may appear as two separated belts with either smooth or subtly ragged edges. Details inside any of Saturn's Rings never appear quite the same from one observation to the next. Keep an eye out for any slightly brighter or darker condensations that may appear inside the B Ring. In short, there are many different, although very subtle, details that can be glimpsed anywhere in Saturn's disk or rings.

Figure 4.1.1 starts with a very prominent feature that usually appears on Saturn's Rings. This is the shadow cast on the rings by Saturn's globe and will appear different in thickness and position each time you observe it. Here, I used a 2B pencil since this feature is very dark.

Figure 4.1.2 shows the rendering of the shadow produced by Saturn's rings on the globe itself also using a 2B pencil.

Figure 4.1.3 illustrates the blending of this shadow with my finger.

In **Figure 4.1.4**, I switch to a 4H grade of pencil. Using a light touch, you can begin to render slightly darker B Ring features.

Figure 4.1.5 shows the B Ring features being softened using a clean blending stump. You should take your time with this since these features are subtle.

Figure 4.1.6 shows the beginning of discerning the placement of the Southern Equatorial Belt. You can start by making a light mark to locate where this belt is positioned in the center on Saturn's disk. For this, I continue to use a 4H pencil.

Figures 4.1.7 and **4.1.8** illustrate the rendering of the Southern Equatorial belt in full. Many times, the Southern Equatorial Belt can appear split with a light-toned rift dividing the northern and southern darker belts. Additionally, the northern belt usually appears thicker than the southern one. From this point on, the positions of any of Saturn's belts can be determined from its center position to its limb position.

You can now blend and smooth the appearance of the Southern Equatorial Belt by using your finger. An eraser shield will be handy in keeping the blending contained to the edge of the Southern Equatorial Belt. This will soften the feature's appearance. (**Figure 4.1.9**)

Figure 4.1.10. It often appears that the entire portion of Saturn's disk, starting from the Southern Equatorial Belt moving toward the Polar Region, is only slightly darker than the lighter-toned equator and B Ring. Use your 4H pencil to lightly shade this area to establish its appropriate tone. Take your time in render-

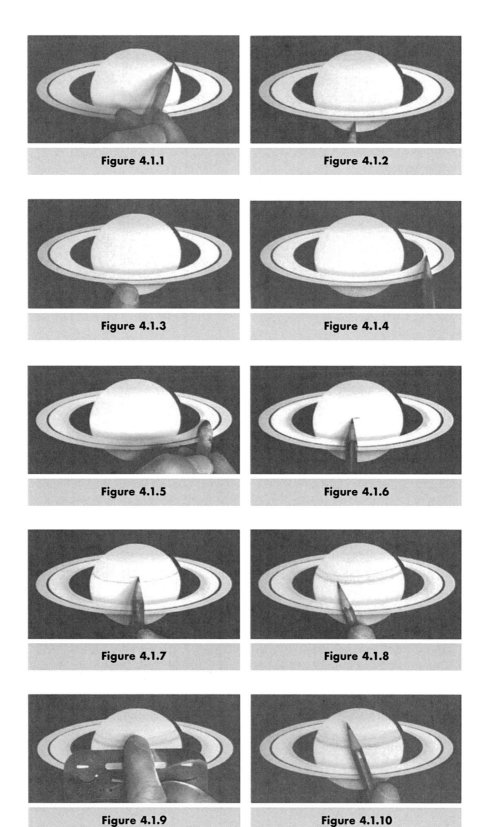

ing the depth of this tone because it is very easy to make the tone too dark. You have to go back and forth, observing through your telescope and comparing what you see, to gradually duplicate this on your sketch.

Figure 4.1.11

Technical Tip

The positions of any of Saturn's belts or the extent of shading in the Polar Region can be determined from the feature's center position to its limb position. Before drawing the entire extent of a belt feature, you can make a light marking that corresponds to how its limb position correlates to the rings, as shown in **Figure 4.1.11**.

The following steps continue by working from the Polar Region back toward the Southern Equatorial Belt. A 4H pencil is still used for adding any new details that you observe and refine any shading that you need to do.

Figure 4.1.12 begins the shading and extent of the darker Polar Cap. You should really spend some time trying to observe any subtle features that may appear in this area. Shading with your pencil should be added in layers. The repetition of shading these layers to the correct amount of darkness is guided by what you observe through your telescope. You can occasionally blend or smooth the shading with your finger, as shown in **Figure 4.1.13**.

Figures 4.1.14 and **4.1.15** show the rendering of thin belts observed in the Polar Region. For these belts, I used the placement method described in **Figure 4.1.6** and the associated technical tip shown in **Figure 4.1.11**.

Very soft blending of these features is achieved by using my finger, as shown in **Figure 4.1.16**. After this blending, another subtle belt was added in **Figure 4.1.17**. As you can see, any time I add another belt feature, I always follow up on it with some blending using my finger, as illustrated in **Figure 4.1.18**.

Figure 4.1.19 shows how I render the southern edge of a major belt feature. This edge always appears subtly mottled to me. I use an eraser that has been cut with a sharp blade to give a pointed, chiseled edge. I simply tap or lightly roll the eraser against the paper at the location where I have observed any raggedness in the edge.

Figure 4.1.20 begins the inclusion of any arc features, known as Intensity Minima, which can be observed in Saturn's B Ring. A very light touch is needed in concert with a 4H pencil to render these. Intensity Minima are usually very subtle and sometimes can be washed out by Saturn's bright light.

Figure 4.1.21 shows the rendering of any dark features that were observed within the A Ring. The feature shown here is called the Encke Minima. This feature appears very different from one observation to the next, if it appears at all. The ability to see any details within any of the rings can sometimes be enhanced with use of the averted vision technique. A light or medium blue Wratten filter also works well for A Ring features.

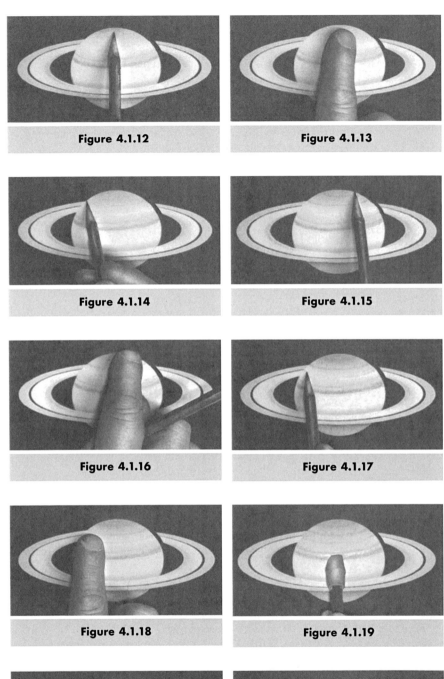

Figure 4.1.12

Figure 4.1.13

Figure 4.1.14

Figure 4.1.15

Figure 4.1.16

Figure 4.1.17

Figure 4.1.18

Figure 4.1.19

Figure 4.1.20

Figure 4.1.21

Observing Tip

At first glance, the Polar Region darkening is relatively easy to observe. However, you may find several kinds of subtle features appearing within this area. Sometimes the Polar Region appears uniformly dark. Other times, there may be a lighter-toned clearing. During other observing sessions, a slightly darker oval-shaped spot appears at the location of Saturn's pole as illustrated below in **Figures 4.1.22, 4.1.23, and 4.1.24**. At times, I have observed both a clearing and the darker spot at the pole. By taking your time while observing, many of these low-contrast features could make themselves visible to you.

Figure 4.1.22

Figure 4.1.23

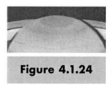

Figure 4.1.24

The following four steps finish up this sketching tutorial. Of course, you can always continue refining your sketch as you compare what you actually see through your telescope with what you have actually sketched.

Figure 4.1.25

Blending Tip

I use an eraser shield for many different tasks. This and the other tutorials I present here illustrate these various uses. Using a finger to blend within Saturn's Polar Region can be pretty clumsy. When executing much of the blending that must occasionally be done, the eraser shield can be used as a blending shield, as shown in **Figure 4.1.25**. This keeps the blending contained so it will not spill over onto Saturn's Rings. You really should not feel constrained by how any of these simple tools are used in this book. They all can be exploited for your benefit and their use for various techniques is wide open to your needs and imagination.

Figure 4.1.26 shows the addition of the very fine Equatorial Belt that divides the light-toned Southern Equatorial Zone and Northern Equatorial Zone.

Next comes the shading that is needed to render the Crepe Ring crossing over Saturn's disk in **Figure 4.1.27**. This is then followed up with any darkening that you may see around the edge or edges of Saturn's limb. For this, you simply use your clean fingertip to lightly rub and smudge the graphite on your sketch to slightly darken these edges.

Figure 4.1.28 shows how to darken the edges of Saturn's globe. This feature is called limb darkening. You can lightly rub a clean finger over the edge since this will pick up a small amount of graphite that is already on the sketch to produce a slightly darker tone.

Sketching the Planets

Figure 4.1.26

Figure 4.1.27

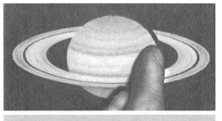
Figure 4.1.28

Figure 4.1.29

Figure 4.1.30

Finally, **Figure 4.1.29** renders the Crepe Ring. I simply use a standard white colored pencil that that you can find in most art stores. It does not take much pressure on this pencil to produce a tone that would be similar to the one that you may observe. Since I used a premade template showing Saturn's outlines printed from my computer, the black background and Cassini's Division are premade.

Figure 4.1.30 shows the finished sketch.

4.2 Sketching Jupiter

Our second sketching tutorial will be of Jupiter, a fascinating and ever-changing world. The marbled details within its belt structures are in constant flux. Light and dark spots appear, and disappear while they make their journey across this planet's disk. Darker rifts can change direction in surprising ways. You should keep a sharp eye focused on ascertaining these features and enjoy the beauty of a particular favorite feature of mine: the wispy smokylike structure of Jupiter's festoons that rise up into the equator. If the Great Red Spot is visible while you are observing, take some time to see if there are any subtle tone changes inside it. Many times, there are small condensationlike details within it. I usually use only a 4H pencil for the middle tones and 2B pencil for the darks. The 4H pencil is being used in all the steps shown to the left.

As shown in **Figure 4.2.1**, you first can indicate the boundaries of the Northern Equatorial Belt and Southern Equatorial Belt by using the straight edge of an eraser shield and a 4H pencil. There is a template provided in the Appendices section of this book for sketching Jupiter. This template indicates the approximate positions for the equatorial belts and was used for this sketching demonstration.

Once you establish this, any dark bumps or condensations can be arranged along the edges of the equatorial belts, as in **Figure 4.2.2**. You can get any relative distances between these first basic features and their proportions by triangulating each feature against any of the others. This technique can be used to proportionally locate these beginning features over Jupiter's disk. You can use this "connect the dots" approach by using any imaginary geometric shapes or angled lines that help you achieve good positioning, seen as relationships between details (illustrated by the red lines). This triangulating of details will become more important as the sketch progresses. The reason for this is that Jupiter's rotation is rapid. As the planet rotates, you will notice that the details shift their position on Jupiter's globe as they move from the following side toward the preceding side. This is also why you may find it to your advantage to concentrate on rendering the preceding side details first.

Your next step is to continue adding basic features that appear throughout the equatorial belts, as in **Figure 4.2.3**. Here, the edges of the belts are more fully established. This sketched example shows the Great Red Spot and features that extend from it. Festoons emerging and any dark features within the Northern Equatorial Belt are also rendered at this point. The gradual shading of Jupiter's Northern Polar Region and Southern Polar Region can be started now. Use the 4H pencil at a shallow angle and rapidly shade to the desired darkness, using very light pressure on the pencil. I usually blend this shading with my finger. Using an eraser shield can also help contain the blending to exactly where I want it.

Figure 4.2.4 shows the use of the eraser shield's straight-edged guide to establish subtler belt features. Other subtle belts observed across Jupiter can be rendered the same way, particularly in the Southern Polar Region. While you are observing, it might be best to see if there are any slightly darker knotlike structures here. Usually the belts in these areas are lumpy looking and do not appear as a continuous straight line across Jupiter's disk

Figures 4.2.5 and **4.2.6** simply shows a bit more progress in adding the Southern Temperate Belts in the Southern Polar Region and dark ribbonlike rifts in the Northern Equatorial Belt.

Sketching the Planets

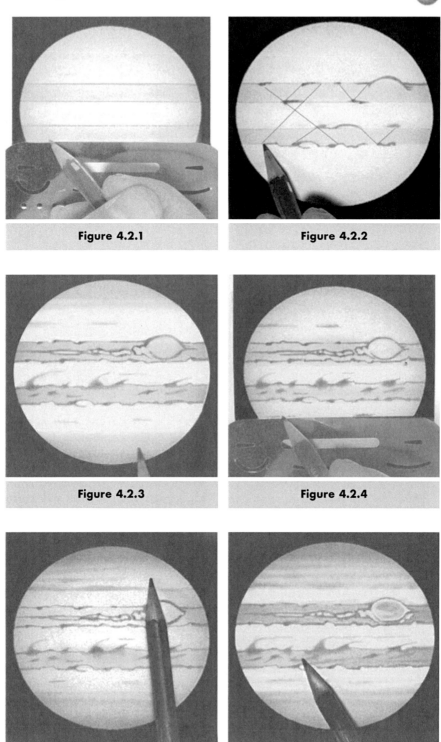

Figure 4.2.1

Figure 4.2.2

Figure 4.2.3

Figure 4.2.4

Figure 4.2.5

Figure 4.2.6

Technical Tips

The eraser shield can also help in containing a blended area's boundary or help smooth out a linear belt feature, as shown in **Figures 4.2.7, 4.2.8,** and **4.2.9**. Other subtle belts observed across Jupiter can be softened the same way. You can also use one of the various blending stumps available at art stores for blending, but I find it quicker and easier to use my finger and an eraser shield.

Figure 4.2.7

Figure 4.2.8

Figure 4.2.9

Figure 4.2.10 illustrates some intensification of the tonality of one the Northern Temperate Belts. This is the same process that was used for the Southern Temperate Belts.

When observing the Northern Polar Region, keep an eye out for any subtle banding or brightening that may be present. In this case, there is some low-contrast banding as well as some progressive darkening of the Northern Polar Region approaching its most northward areas as shown in **Figure 4.2.11**. You can follow up after you have shaded these in with blending to soften the gradations.

Figures 4.2.12 and **4.2.13** illustrate the details observed and rendered within the Southern Equatorial Belt and Great Red Spot. Here, I switch to a 2B pencil. Switching to the 2B helps in applying a darker tonality with more speed. At this point, you should concentrate on the intensity of any boundaries and dark condensations that you have observed. It would be best for a gradual buildup from thin to thick in any of your linear renderings. It is simply easier to add to this part of your sketch as opposed to subtracting by erasing. That said, any light spots could be added later if necessary. These details can be added in any order since details vary as you observe them.

Figure 4.2.14 focuses on getting the festoons rendered with more articulation. Observing festoons can be difficult since the comparatively bright Equatorial Zone usually surrounds them. Once you settle the full extent of each festoon as it reaches out into the Equatorial Zone, you can then render how they interface on their northern edges within the comparatively Northern Equatorial Belt. At this point, you may wish to switch back and forth between the harder graphite of the 4H pencil and the softer, darker graphite of a 2B pencil. **Figure 4.2.15** illustrates this and also shows the use of an eraser shield to contain any blending you may need to do within a boundary.

Sketching the Planets

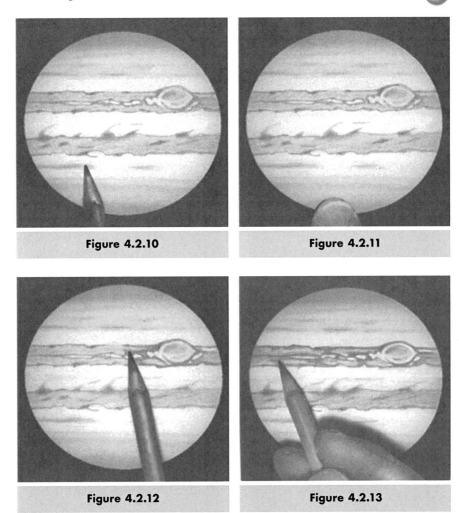

Figure 4.2.10

Figure 4.2.11

Figure 4.2.12

Figure 4.2.13

Figure 4.2.14

Figure 4.2.15

At this point I would like to digress a bit . . .

All of the observing tricks that you may use while observing really start to come into play while trying to sort out the various details and changes in contrast within any of the belts and zones that you may encounter. Averted vision plays a major role as a technique I use for catching lots of little details. I do not really get into the distinctions between different celestial object types while observing. To my mind, a planet is simply a very bright type of nebula.

My experience is that when I get to this stage of a planetary observing session, all sorts of things start to appear at any point and at any moment within a planet's disk. You may find it every bit as confusing and overwhelming as I did when I started sketching. If you are like I was, you might find that your sketches never really get fully completed. I used to just concentrate and make a very small drawing of one of Jupiter's belts with dimensions of say 2″ across by 0.25″ high. An example of this is shown in **Figure 4.2.16,** which is a strip rendering of the Southern Equatorial Belt. **Figure 4.2.17** is the Northern Equatorial Belt. This is also similar to the scale of the equatorial belts provided in the Jupiter drawing template in this book.

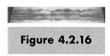

Figure 4.2.16

Figure 4.2.17

Your observing skills improve when you see details inside of details. When this happens while sketching, it is the more magical part of observing. If you are a beginner at sketching, do not get frustrated by not being able to finish a sketch that matches either your expectations or your ability to render what you have seen. In my opinion, all the above are actually the major points for taking the time to sketch while observing.

In time, your sketches, and the ability to use them as a form to communicate and record what you have observed, improve. While your sketches improve, so does your ability to observe and sort out details. As time passes, your ability to abbreviate all of this becomes easier and faster.

Figure 4.2.18 begins to add the final details and adjust the tonal contrast balances within the Northern Equatorial Belt. If you need to broaden a lighter mass tone in the Northern Equatorial Belt, tap, or very lightly roll, your eraser over those areas. The idea here is to lift the graphite off the paper rather than rub it off. The curved, cut-out holes on the eraser shield help contain the eraser to a basic shape, which can then be followed up using a 4H pencil for more accuracy.

Figure 4.2.19 reestablishes some darks, but in this particular case, I am rendering a lighter oval feature that I happened to observe. You can start by first using one of the smaller dotlike cut-out holes on the eraser shield to erase the smaller spot. Then follow up by using a 4H pencil to accurately render this feature's shape and outline.

Figures 4.2.20 and **4.2.21** finalize this process.

Figures 4.2.22 and **4.2.23** show the rendering of two white ovals in the Southern Temperate Belt. These white ovals can be very difficult to observe. A telltale sign of their appearance is that you may see something of a dark condensation surrounding them.

Sketching the Planets

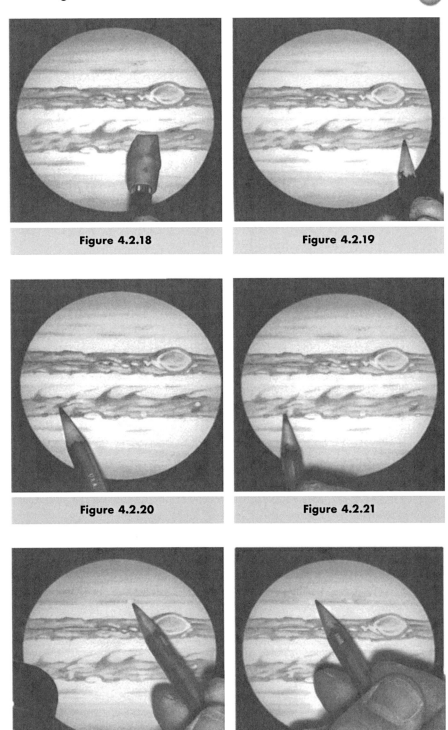

Figure 4.2.18

Figure 4.2.19

Figure 4.2.20

Figure 4.2.21

Figure 4.2.22

Figure 4.2.23

Technical Tip

An eraser shield can be very helpful in removing graphite in very small areas. It would benefit you greatly to utilize every type of cut-out shape. Below are some illustrated erasing steps that were used in rendering the Northern Equatorial Belt.

  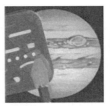 

| Figure 4.2.24 | Figure 4.2.25 | Figure 4.2.26 | Figure 4.2.27 |

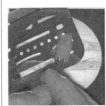 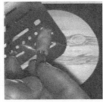

| Figure 4.2.28 | Figure 4.2.29 |

Technical Tip

Observed White Ovals can be rendered in the following manner, shown in **Figures 4.2.28** and **4.2.29**. Using the small dotlike cut-out holes on the eraser shield, along with an eraser, will help indicate where they are. You can follow up with very light pressure using a 4H pencil to establish their actual size, taking care not to render them too large.

NOTE: You can use your pencil's point, held just above the drawing paper, to position the eraser shield. The eraser shield can be slid under it for accurate placement before erasing.

These are last steps of this tutorial. Now is the time to simply make adjustments to the previously rendered details as well as adjustments in tonality depth. **Figures 4.2.30** and **4.2.31** show some accentuation in the Northern Polar Region's subtler belts and the darkening of the most northward region. Again, I use my finger to blend and soften the appearance of these details.

Figure 4.2.32 shows the rendering of any limb darkening on Jupiter's disk. This limb darkening usually, but not always, makes itself visible. Additionally, the degree of darkness on limb changes quite bit from one observation to the next. For this, I gently rub a clean fingertip along the edges of the sketch.

Figure 4.2.33 shows the finished sketch.

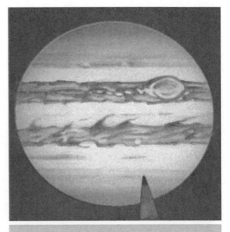

Figure 4.2.30

Sketching the Planets

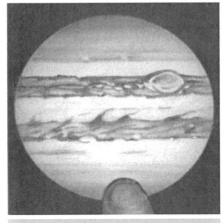

Figure 4.2.31

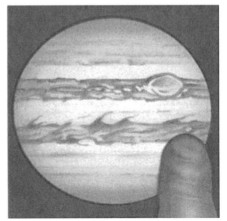

Figure 4.2.32

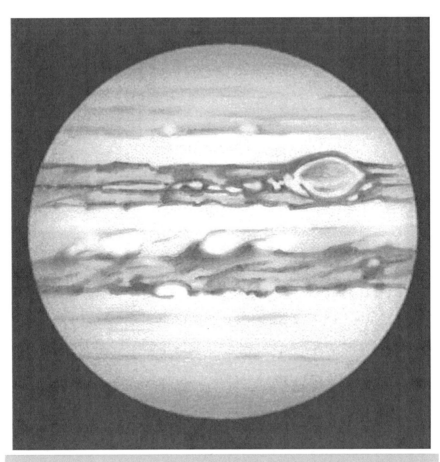

Figure 4.2.33

4.3 Sketching Mars

This last tutorial shows the various sketching and erasing techniques I use to render Mars. Sketching Mars is much more complicated due to the wide range of features and the tonal contrasts that exist between them. Mars has a rapid rotation rate, so its features change their positions quickly. You might want to spend some time observing Mars to sort out various details before you begin sketching. I typically observe between 30 minutes to an hour before I begin a sketch.

Every step in this tutorial is illustrated to show how I do a complete sketch. This particular example shows Mars at opposition, so its silhouette is a circle. At other times, Mars will show a "phased" appearance similar to the phases of the Moon. During these times, you can simply color in the appropriate silhouette to achieve a phased outline of Mars' disk.

Figure 4.3.1 starts with the rendering the edge of the Southern Polar Cap. For this, I use a 4H pencil. You should use a very light touch and a very small semi-circular motion to achieve a desired thickness that matches up with what you observe through your eyepiece.

What follows are the very important steps that will establish a base tone over the rest of Mars' disk.

Figure 4.3.2 shows me using my 4H pencil to lightly tone everything within the disk except for the white Polar Cap.

Figure 4.3.3 shows blending being done to smooth out this tone. I use my fingers. It is best to take your time and repeat these two steps to gradually build up this tone smoothly until you reach the desired shade.

> **Technical Tip**
>
> Shading a large area to get a uniform tone is best accomplished by holding your pencil at a very shallow angle and using long, straight stokes. This way, you use the side of the graphite rather than the sharpened point. It also requires applying very light pressure on the pencil. You will find it advantageous to randomly change directions when shading each time you need to repeat these steps. I use my finger to blend and smooth out these strokes each time after I shade with the pencil. Here I find it best to use a circular motion and favor blending in random directions.

Figure 4.3.4 illustrates the beginning stages for rendering the darker albedo features, still using the 4H pencil. Holding your pencil at a shallow angle to your drawing surface will benefit you in controlling your pencil's position and pressure. You should continue to use very small, semicircular stokes. Throughout the rest of the steps presented here, the focus will only be on the darker albedo features.

Figure 4.3.5 continues with the addition of other dark albedo features.

Figure 4.3.6 again shows me blending with my fingers. This will automatically darken and smooth out very large areas of Mars' globe. I blend every time I add a new portion of an albedo feature.

Sketching the Planets

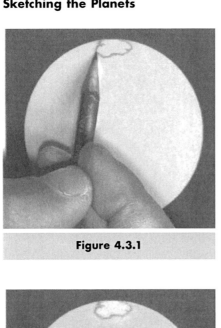

Figure 4.3.1

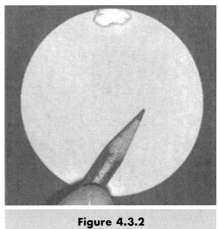

Figure 4.3.2

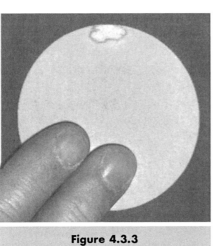

Figure 4.3.3

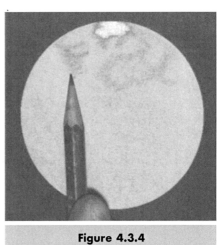

Figure 4.3.4

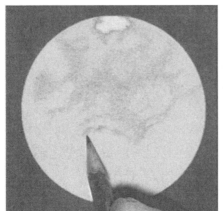

Figure 4.3.5

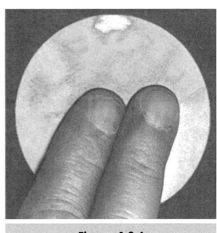

Figure 4.3.6

Figures 4.3.7 and 4.3.8 continue the same process for the remaining dark albedo features as described in the previous steps. At this point, all of the major dark features that I have observed have been roughly sketched. If you are a beginner and have made it this far, you have accomplished a lot. Congratulations!

Technical Tip

There are many valid approaches to start a sketch of Mars' darker albedo features. One approach is to draw an outline of the darker features and follow up by filling them in with any details you are able to observe. If you use this approach, I suggest that you render the outlines first and then follow the rest of the steps as outlined in this tutorial.

That said, I sketch in the opposite manner and begin sketching from the inside of these features and move outward.

Comparisons Between Observing and Sketching

At this point in the tutorial, I want to explain my opinions about observing and how it relates to sketching. Now that the major albedo features of Mars are roughly indicated in your sketch, I would like you to consider how observing and sketching can help you glean greater rewards and personal satisfaction from our hobby.

It has often been said that sketching can make a better observer. I believe one reason for this is that you can compare details you are able to observe against the details you can sketch. This comparison can hone your abilities to sort out any differences between what you notice between the observed and sketch details. During the period of time it takes for you to execute your sketch, these two kinds of information can combine and become layered. You will find that as you keep refining your sketch, the areas that you work on and observe get smaller and smaller. This will happen regardless of what your capability as a sketch "artist" is.

At this stage of the tutorial, you can begin the process of being able to observe any finer details inside the details you have already recorded. For sketching, this amounts to recording details as shapes within shapes. This also applies to comparing the depths or degrees of light and dark tones within any shapes in the details you see. As you turn your attention to this deeper level of detail, sketching only seems to become a bit more complex. Simply speaking, this is a process or skill that I believe helps you become a better observer as well as a sketcher.

Mars, more than any other observable planet, is replete with all kinds of detail that make the comparisons I mention a real challenge. That is why this tutorial for sketching Mars is the most complicated, because it has many details appearing all over its surface. Any success or shortfall you experience observing and sketching Mars will end up being a great benefit when you turn your attention to Saturn, Jupiter, or any other celestial object that you choose to sketch.

Sketching the Planets

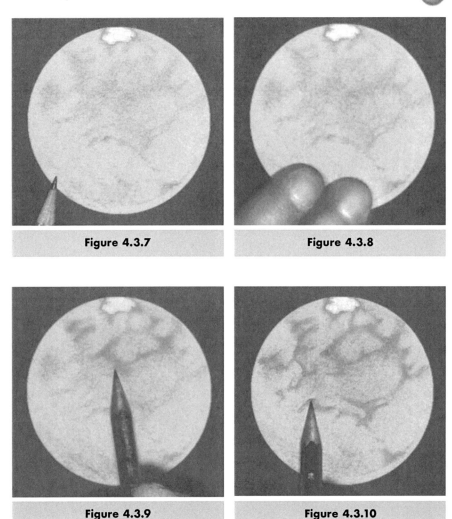

Figure 4.3.7

Figure 4.3.8

Figure 4.3.9

Figure 4.3.10

Throughout the following steps you will also make any necessary corrections to the boundaries and proportions of the darker albedo features that you have indicated thus far. Use the same methods and 4H pencil that you have been using from the start of this tutorial. At this point, you concentrate only on the darkest features. There is no blending throughout the next four steps shown in **Figures 4.3.9** through **4.3.12**.

Figure 4.3.9 starts the process of reestablishing and rendering the darkest albedo features with more accuracy. Using a 4H pencil and starting at the Southern Polar Cap, you can hone in on its dark collar. From this point onward, you should look for and reinforce any boundaries and condensationlike features that appear as a darker tonality.

Figure 4.3.10 shows details being added as we move southward on the sketch.

Figures **4.3.11** and **4.3.12** show more details being added to this sketch by more accurately reestablishing a network of the darkest albedo features. This finishes the process of redefining the darkest albedo features.

> ## Technical Tip
> Confining the blending to a smaller area will be very helpful to you from this point forward. You will notice that after blending, the middle tones for the dark albedo features appear a bit darker than before. You can see this by comparing the differences between **Figure 4.3.12** and **Figure 4.3.13**.

Figure 4.3.13 shows finger blending using a very light touch. The idea here is only to slightly soften what has been rendered in previous stages of your sketch. You should confine this blending over only the dark features, leaving the lighter toned areas untouched.

Figures 4.3.14, **4.3.15**, and **4.3.16** repeat the process of refining the darkest of the albedo features. For this, I have switched to a softer 2B pencil, using extremely light pressure and the same shallow angle to the drawing surface that has been used throughout this tutorial. You should maintain your concentration while observing, and sketch only the dark features.

This completes the rendering of the darkest albedo features.

> ## Technical Tip
> The rendering of the lighter albedo features and tonal corrections for lightening any medium toned features will be accomplished by using an eraser and an eraser shield. You can use a sharp blade to cut an eraser so it has a sharp chiseled edge and sharply pointed corners. Doing this helps you to achieve more control when erasing.
>
> Most erasing only requires lifting a little bit of graphite from an area to lighten a tone.
>
> Either tapping the eraser or lightly rolling it over the area of your sketch that needs to be lighter in tone can do this. Generally speaking, tapping the eraser on the drawing surface lifts a little bit more graphite and is better suited for smaller areas than rolling the eraser.
>
> If you lighten an area too much, you can very lightly rub your finger over this area to darken it. Any boundaries between lighter and darker areas can also be reestablished by using a 4H pencil.
>
> Rubbing an eraser back and forth the traditional way is done only to achieve the brightest white tones.
>
> Throughout your observation and sketch of Mars, you have probably observed many lighter toned or very bright albedo features and their relative positions. Take your time to render these features now.
>
> From this point forward, it would be a very good idea to carefully compare each successive stage shown in this tutorial to see the differences that erasing makes. The differences in these steps range from subtle to striking.

Sketching the Planets

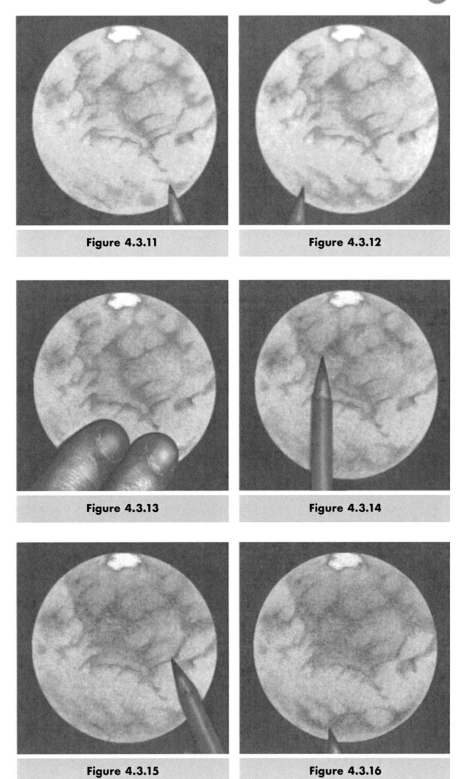

Figure 4.3.11

Figure 4.3.12

Figure 4.3.13

Figure 4.3.14

Figure 4.3.15

Figure 4.3.16

Figure 4.3.17 illustrates some preparatory light pressured finger blending to smooth out just the lighter areas. Now you can now turn your attention to the lighter toned features you have probably noticed throughout your observing session.

Figure 4.3.18 shows the lifting of graphite by tapping the corner of the eraser lightly on the drawing paper's surface. Using a sharply pointed edge of your eraser offers you a lot of control by lifting graphite out of very small dotlike areas. Repeated tapping from the center of an area outward allows for a graduated lifting of graphite. This process can be controlled to remove less and less graphite as you reach the outer boundary of the area that you need to lighten.

Figure 4.3.19 shows the same technique being used in another area. As this area is a bit larger, you might want to use a combination of tapping and rolling your eraser on the paper's surface.

Figure 4.3.20 illustrates the use of a 4H pencil for correcting and reestablishing the edge of a feature that was erased a little too much. Small corrections like this may need to be done from time to time as you fine-tune your sketch.

Figure 4.3.21 shows the effect of using both techniques—rolling and tapping an eraser—to lighten a large area. I use the eraser shield to cover and protect an area so I do not make this lighter area too large. I move the eraser shield as I progress over the area that needs to be lighter in tonality. Again, I prefer to work my way from the area's center outward to its edge.

Figures 4.3.22 and **4.3.23** show the lightening up of the smallest areas. Here is where the precut shapes in the eraser shield really help in lifting graphite that is close to the edge of a darker area's boundary. This is to protect any of the veinlike dark albedo features that you have previously sketched. Here, you either roll or tap the eraser over this cut out shape. Just be careful that it is accurately positioned over the area that you wish to lighten.

Figure 4.3.24 continues the erasing process in a different area on Mars' globe.

As mentioned in the technical tip in the table on page 92, you can see the differences that these steps have made in rendering the depth of the various contrasting tonalities found in the albedo features. The range of these differences that can be seen in **Figures 4.3.17** and **4.3.24** is quite dramatic.

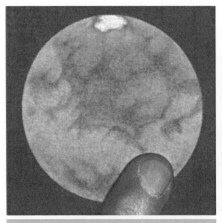

Figure 4.3.17

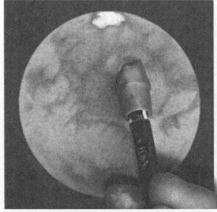

Figure 4.3.18

Sketching the Planets

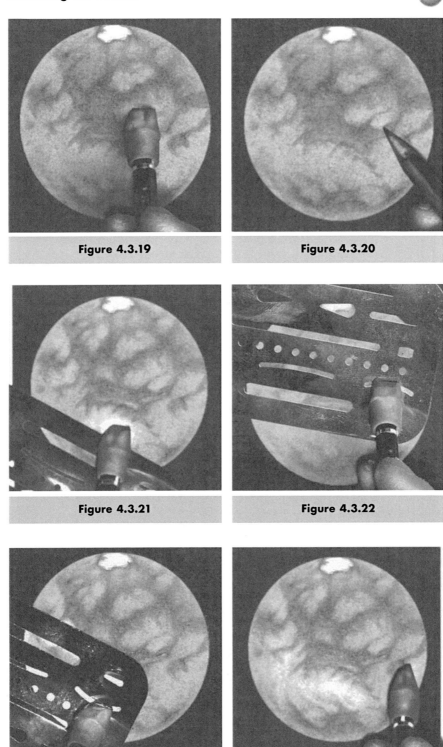

Figure 4.3.19

Figure 4.3.20

Figure 4.3.21

Figure 4.3.22

Figure 4.3.23

Figure 4.3.24

Figure 4.3.25 shows a slightly different use of the eraser shield. Note here that sometimes it may be difficult to see just exactly where the opening of the shield is placed. I am now using the widest opening to achieve some very delicate and precisely located lifting of graphite.

> ### Technical Tip
> What follows are the final steps of this tutorial.
>   You will now use the eraser to show the lightest details that can be observed on Mars. These are usually cloud or dust features that can be seen around the edges of Mars' globe.
>   You can now use your eraser in the traditional way by rubbing it back and forth. Your goal is simply to remove all of the graphite and achieve the whitest tone possible. It is best to make sure to use a part of your eraser that still has a sharp chiseled edge.

Figure 4.3.26 shows the rendering of what is known as the Northern Polar Hood. You should still strive to be accurate to the size and extent of this area as well as any other brightened area that you wish to render. In this case, I find it best to slowly work my way from the outer edge of the disk toward the inward boundary line of any given bright area. At this stage, it is very difficult to go back and make corrections if you erase too far into Mars' disk; so take your time.

Figure 4.3.27 continues the traditional use of the eraser to show the bright edge of Mars that is known as limb brightening. The eraser's position shown in this step is at the location of a bright bulge associated with a cloud or haze. This brightening is caused by sunlight that is scattered onto and through the thicker parts of Mars' atmosphere. Features like these can be observed along the Martian sunset or sunrise.

Figure 4.3.28 finishes this tutorial with the rendering of limb brightening along Mars' other side.

Figure 4.3.29 is the finished sketch.

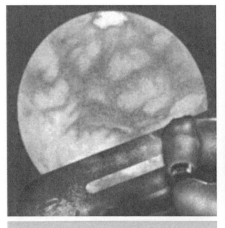

Figure 4.3.25

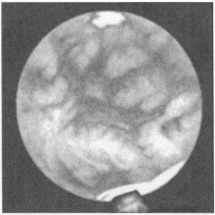

Figure 4.3.26

Sketching the Planets

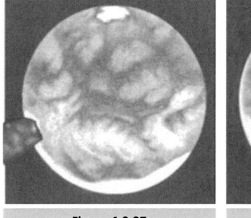

Figure 4.3.27

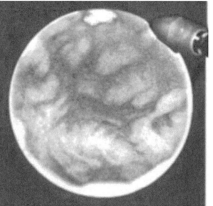

Figure 4.3.28

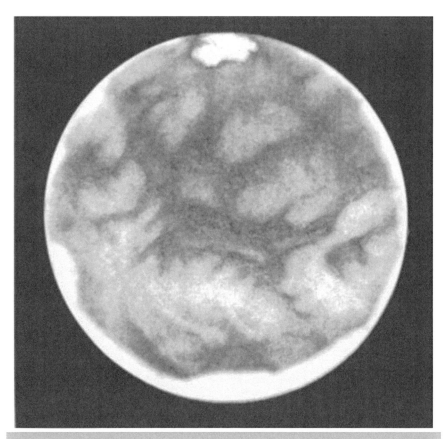

Figure 4.3.29

CHAPTER FIVE

Sketching Star Clusters

The next time you plan a quiet evening under a salted sky, with hopes of bathing your eyes in the ancient light of a majestic star cluster, be sure that your sketching kit comes with you! A casual glance at these celestial marvels will not give you a decent appreciation for an object whose history and character are as unique as the fingerprints you should be pressing into the side of your trusty pencil. I can think of no better way to connect with these stellar ballets, to understand their intricacies, and to recall your view later than to spend time sketching the soft glow or blazing pinpricks you see through the eyepiece.

Whatever your motive for sketching these distant spectacles, you may find yourself stumped by the complex clumps, swirls, and streamers of stars revealed through your eyepiece. Although some of the larger, more complex clusters can prove to be very rewarding subjects, it is a good idea to start simple while honing your sketching technique. Rather than beginning with something like M11 (the Wild Duck Cluster) or NGC 5139 (Omega Centauri), start with objects that are a bit more manageable. As you gain confidence with these simpler studies, you will be better prepared to move on to more challenging objects. The tutorials for open and globular clusters in this chapter will each cover a simple and complex object. This will help you to get started and then provide material for moving on to more ambitious targets when you are ready.

Keep in mind as you go through these tutorials that you will eventually develop your own sketching style. The techniques discussed here are just a starting point, so do not hesitate to experiment. As you examine a variety of approaches, you will find certain techniques that you do not care to use and others that you want to develop further. You may even develop a technique that no one else has thought of, but works perfectly for you. Whatever methods you use, the act of sketching will help you become a better observer and sharpen your connection to the vast cosmos above.

The following tutorials describe graphite sketching techniques. At its simplest, a graphite sketch can be produced with nothing more than a pencil, paper, a clipboard, and a red flashlight. However, to provide you with more control over your sketch, I suggest the following list of materials:

- Clipboard
- Dimmable red observing light
- Paper prepared in any of the following ways:
  - Blank
  - Prepared with predrawn sketching circles
  - Copied or preprinted log sheets
- HB, 2H, and 4H pencils
- Pen (for notes)
- Blending stump or tortillon
- Choice of erasers (Art Gum, eraser pencil, kneaded)
- Eraser shield (used to constrain erasures to a small area)
- Pencil sharpener or lead pointer
- Sandpaper block (used to hone the point of a pencil, blending stump, or tortillon)
- Small paint brush (used to brush away loose graphite or eraser debris)

Open Clusters

Open clusters provide a dazzling array of forms to view and sketch, from naked-eye wonders such as the Hyades, to binocular delights such as the Double Cluster in Perseus and Kappa Crucis, to any of a vast multitude of telescopic clusters. Although the tutorials in this section deal with telescopic sketches, these techniques can be adapted to work with naked-eye or binocular observations.

I believe that the skills used in sketching open clusters provide a great foundation for sketching other deep sky objects. No matter what you are sketching, whether globular clusters, nebulae, galaxies, or naked-eye vistas, you will almost always lay down a field of stars that surround those objects. That is what open clusters really boil down to: fields of stars, ranging from simple to complex. You will learn to see these collections of stars as networks of geometric shapes that can be translated to a sketch. As you develop this skill, you will be able to take it with you as you record other deep sky marvels.

Before you dive into a sketch, take time to absorb the view. Try to get an initial impression for what makes this collection of stars unique. Does anything jump out at you right away? Perhaps a pattern of stars emerges, reminding you of something distinctly nonstellar. Or you may be impressed by a glint of color in a few stars sprinkled throughout. You might also detect hazy or granular patches that hint at masses of unresolved stars in the background. Take note of these

Sketching Star Clusters

things and jot them down. These notes can prove very helpful as you develop your sketch. The more time you spend on the cluster, the more details will appear. As you focus on these details, some of those initial patterns, colors, or other impressions may dissipate in your mind. Referring to those initial notes will help you be sure to capture as many facets as possible in your sketch.

5.1 Sketching a Simple Open Cluster

To start with, we will take a look at M29, a small, lightly populated cluster in the constellation Cygnus. It will provide a nice starting point for discussing techniques that you can use to bring these stars down onto paper. For this sketch, I used a 15-cm, f/8 Newtonian reflector, with a 10-mm Plössl eyepiece that provides a magnification of 120× and a true field of view of 24 arc minutes. The scale of your sketch and the number of visible stars that you are able to plot will vary, depending on the telescope and eyepiece combination that you use, as well as the quality of the sky where you observe.

This sketch was made under mild light pollution, and as a result, true dark adaptation was not possible. Keep that in mind as you see this sketch develop. If you are observing with a similar telescope from a very dark site and are nicely dark adapted, you may pick up many more stars in and around the same cluster. Conversely, if you are observing from a location with heavier light pollution, you may make out only a few of the brightest stars.

I located the cluster using a low-power eyepiece and noticed that it did not dramatically pop out of the glittering Milky Way backdrop like many other Messier clusters do. It may be tempting to consider such a cluster worthy of only a passing gaze. However, I find it rewarding to look for the unique qualities that give these more humble objects a personality unique from any other. Immediately, this little cluster displayed a geometric, almost artificial appearance to me. Some have described it as a little teapot, a small dipper, or a miniature version of the Pleiades. These are all great analogies, and to those, I added my own gut impression of a miniature circuit board plugged into the heart of the Cygnus Milky Way.

Step 1: Preparing the view and sketch area Having spent time observing the cluster, adjust your telescope if necessary to frame it properly for the sketch. If you have room for adjustment, try placing an easily identifiable star at the center of the eyepiece. This can be very helpful in estimating the position of other stars as you work your way through the sketch.

Adjust the brightness of your red light to the dimmest setting that you can manage. Of course, you want to be able to see what you are sketching, but as your eyes become adjusted to the dark, a fainter light will be easier to use, and you will be able to see much more through the eyepiece. For additional information on balancing dark adaptation with the use of a sketching light, see *Sketching Faint Objects in Low Light* on page 42.

The first step to take with the sketch is to determine cardinal directions and mark them outside the sketching circle (see *Assessing Cardinal Directions for Your Sketch* on page 39). Having this marked will help with note taking if you

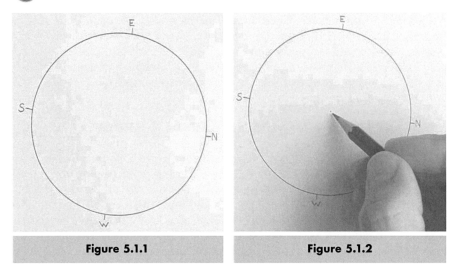

Figure 5.1.1   Figure 5.1.2

need to describe the position of a star, structure, or passing satellite during the course of the observation. (**Figure 5.1.1**)

Step 2: Plotting bright framework stars

If you were able to center your view on a notable star, carefully position your pencil at the center of your sketch and mark this star first. (**Figure 5.1.2**) For tips on marking stars in your sketch, refer to *Marking Stars* on page 126. I prefer to initially plot all stars with a 2H pencil since the marks are light, easily controlled, and can be corrected more readily if necessary.

Next, mark the brightest stars in the view to form a framework for the rest of your sketch. This framework will be very important to keep your developing sketch proportional throughout, so be as careful as possible with your star placement at this stage. Try visualizing a clock face or an X whose arms rest at 45 degrees, along with concentric circles as you look through the eyepiece. Use this imaginary structure to estimate where these stars should be marked.

Begin by marking the brighter stars that come closest to an obvious angle and distance from the center. Unfortunately, this view of M29 did not provide many simple alignments. In **Figure 5.1.3**, you will notice that one star (A) was in the 10 o'clock position about seven-eighths of the way to the field stop. Another star (B) was just shy of the 7 o'clock position and five-eighths of the way to the edge. Continue this process, taking care to mark the position of these stars as accurately as possible.

As you add more stars that do not relate easily to the imaginary lines visualized in your mind, start making comparisons to other stars you have plotted. For example, the star marked D in **Figure 5.1.3** was just shy of the 4 o'clock position and about the same distance from the edge of the view as the star marked C. It was also about as far above B as B was above C. The more you look for multiple relationships that relate to a stars position, the better your framework will be.

Sketching Star Clusters

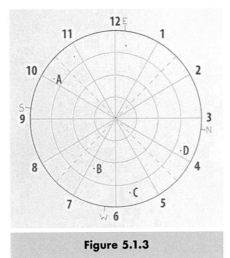

Figure 5.1.3

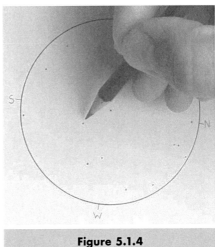

Figure 5.1.4

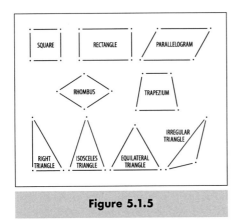

Figure 5.1.5

Try to work from larger- to smaller-scale arrangements, becoming more detailed as you add to the sketch. Feel free to use your imagination, thinking of other ideas that will help you relate the position of one star to another. The extra effort you put forth to be accurate here will pay off later as you fill in the other stars, using these framework stars as a reference. If you make a mistake in placing a star, take a moment to either erase and replot it, or mark it for correction later (see *Correcting Misplotted Stars* on page 128).

Step 3: Plotting bright stars in the cluster Now you are ready to add the brighter stars within the confines of the cluster itself. With M29, I decided to start with a bright star that bounded its southwestern corner. This star and the central star of the sketch bracket two sides of the cluster and serve to place the remaining stars proportionately. (**Figure 5.1.4**)

As you begin to concentrate on stars that are in closer proximity to each other, it becomes easier to find linear arrangements, geometric shapes, and close pairings. Try to find squares, rectangles, parallelograms, rhomboids, trapezoids, and all varieties of triangle: right, isosceles, equilateral, and irregular. (**Figure 5.1.5**) Look for stars that seem to fall along a line between two other stars that you have already drawn. Within open clusters, you will often see double stars scattered about, and you will want to look for and capture these in your sketch as well. Getting the right proportions is still important for any of these relationships.

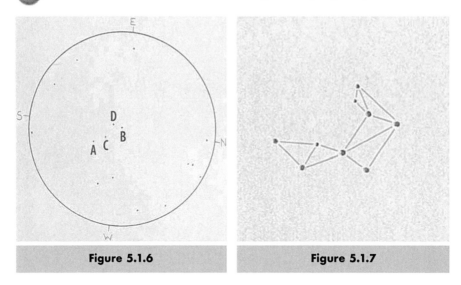

Figure 5.1.6 Figure 5.1.7

Compare these off-the-cuff asterisms against stars you have already sketched to ensure that their sizes and angles are correct.

Using the two bracketing stars of M29 for reference (see stars A and B in **Figure 5.1.6**), I marked another bright star (C) that rested on an imaginary line between the two. Star C was about 40% of the way from A to B. A fourth star (D) formed a slightly obtuse right triangle with stars B and C. **Figure 5.1.7** shows some of the additional triangular arrangements that I used to complete the brighter stars in the cluster.

Step 4: Plotting faint stars in the sketch With the framework of bright stars in place, begin to mark the remaining stars. This is a good time to dim your sketching light further, if possible, to help you better see the fainter stars. As you decide where to begin this part of the process, it helps to be systematic in your approach. One method is to find a starting point at the periphery of the sketch and work your way clockwise or counterclockwise until you are finished. Or you may choose to start with the cluster itself and wind your way outward.

In this case, I started with the remaining stars in the vicinity of M29. (**Figure 5.1.8**) Once the body of the cluster was complete, I moved on to the rest of the sketch. Continue the systematic approach so that you do not miss anything. In this example, I worked counterclockwise from the 12 o'clock position, filling in stars while continually looking for shapes, lines, and angles to help position them. (**Figure 5.1.9**) As you do this, continue working from larger relationships to smaller ones, since this helps keep your proportions accurate.

Step 5: Finishing the sketch At this stage, reassess whether the weight of the stars in your sketch are reasonably accurate compared to the view through the eyepiece. After you determine which stars are brightest, return to your sketch

Sketching Star Clusters

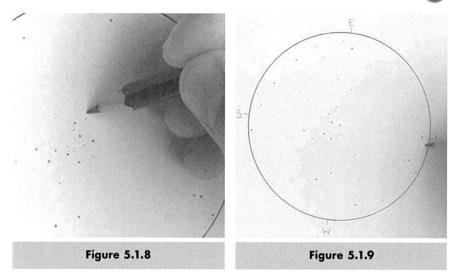

Figure 5.1.8

Figure 5.1.9

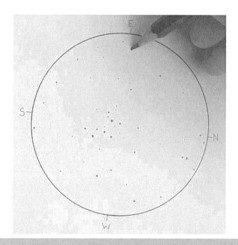

Figure 5.1.10

and see if things match up accordingly. You may wish to switch to an HB pencil at this point, to darken the brighter stars. I enjoy this tidying process, because as the stars get bulked up to their appropriate weights, the sketch starts to take on a very convincing depth. **(Figure 5.1.10)**

With these last items taken care of, it is time to deal with a little red tape. Be sure to note details about your observing equipment, conditions, and the date and time on your log sheet along with any notes that you want to record. The fact that a skunk visited you in the middle of your observation will not show up in the sketch (hopefully), but might be amusing to recall later from your notes. The

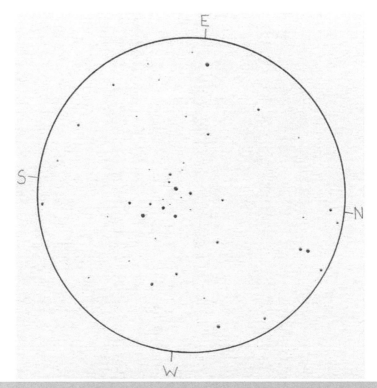

Figure 5.1.11

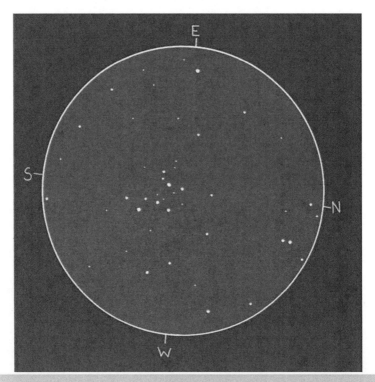

Figure 5.1.12

Sketching Star Clusters

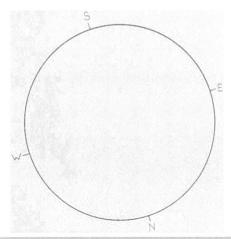

Figure 5.2.1

finished sketch can be seen in **Figure 5.1.11**; **Figure 5.1.12** shows the sketch inverted to white on black.

5.2 Sketching a Complex Open Cluster with Unresolved Stars

Not all open clusters are as simple to sketch as M29. If you find yourself face-to-face with a rich open cluster whose brighter stars rest on a foaming bed of grainy or hazy background stars, you have the opportunity to add some new techniques to your sketching tool belt. In the following section, we will take a look at the open cluster M50 in Monoceros. This cluster is populated with numerous faint stars that embrace its brighter stars in an underlying, almost granular cloud.

As I spent time observing it, I felt that it would be important to include that subtle, inner glow in the sketch. A few other features also caught my attention, including a couple of orange stars and streamers of stars that added a spiraling appearance to the eastern side of the cluster. Finally, the cluster itself rested in a rich Milky Way background with many distinct but faint stars. After making a mental or written note of features such as these, you are ready to begin the sketch.

Step 1: Preparing the view and sketch area In addition to adjusting the view of the cluster so that it is well framed, see if you can place a conspicuous star in the center. In a sketch that is rich with stars, this will make accurate star placement much easier. This sketch of M50 offered a number of bright candidates, and I chose one just east of the hub of the cluster on which to center. With the framing taken care of, take a moment to mark the cardinal directions around your sketch circle. (**Figure 5.2.1**)

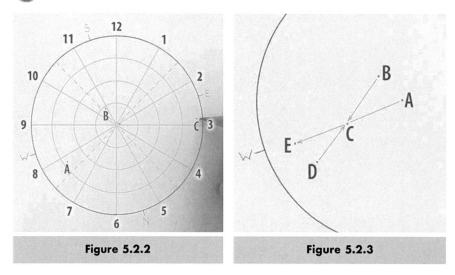

Figure 5.2.2   Figure 5.2.3

Step 2: Plotting the bright framework stars If you were able to place a star at the center of the view, mark that first. As in the previous tutorial, using a 2H pencil will help you keep star weights controlled and conservative at this point. Try to imagine a clock face, 45 degree angles, and concentric circles superimposed in your eyepiece and on your sketch. For example, in this sketch (**Figure 5.2.2**), the star marked A rested a little below 8 o'clock, almost three-fourths of the way to the edge. Another star marked B was a bit below 11 o'clock, about one-fourth of the way to the edge. A third star, C, hovered just above the 3 o'clock position, almost to the edge of the field stop.

With a star field that is heavy with stars, another technique you may find helpful is to slightly defocus your eyepiece. This can help to remove the "clutter" of fainter stars and allow you to concentrate on the brightest. Give this method a try if you find yourself overwhelmed by the mob of stars before your eyes. A note of caution: Some star fields contain such a full range of stellar magnitudes crowded in small spaces that switching back to a sharp focus can introduce some confusion as to which bright stars you actually marked. If this happens, you may want to concentrate your attention on the crowded area, defocus again to reorient yourself, and then sharpen it back up while paying close attention to the stars that you know you marked.

As you proceed through the placement of bright stars, you will start to run into stars that do not line up as easily with the imaginary template of clocks, 45 degree angles, and concentric circles. For example, in **Figure 5.2.3**, the star marked C fell a bit more than halfway along an imaginary line between the stars marked B and D. After marking this star, another star, marked E seemed to rest along another imaginary line drawn through the stars marked A and C. Its distance from C was about the same as the distance from A to C. Continue to use your imagination as you try to find these relationships, keeping them as accurate as possible. **Figure 5.2.4** shows these framework stars in place.

Sketching Star Clusters

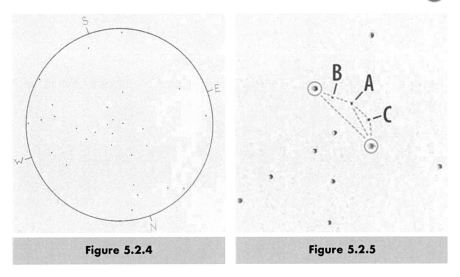

Figure 5.2.4

Figure 5.2.5

Step 3: Plotting stars within the cluster With your framework of bright stars in place, you can begin marking the remaining stars. Give some thought to turning your sketch light down further if you can manage it. You want to allow yourself to see the fullest extent of faint stars possible now. A star field as loaded as this is demands a systematic approach. You could begin working clockwise or counterclockwise from the outside in, or inside out. Which way you choose to attack it is not as important as being deliberate in your approach. If you scatter your efforts, you may soon find yourself uncertain about which stars you have and have not plotted. In this example, I chose to begin at the heart of the cluster and work clockwise.

When you try to render a large number of stars in a very tight space, it can be easy to draw tight groupings of stars larger than they actually are. So it is important to keep a very close eye on your proportions. You may need to exclude the faintest stars in the area if it becomes too crowded. In cramped quarters, you will also need to pay close attention to the position of your pencil tip as it descends on the paper. The margin for error can be pretty slim in these bustling regions. The good news is that as you begin to add more stars, it becomes easier and easier to find simple geometrical relationships to use for placement.

As discussed in Step 3 of the previous tutorial, use linear proportions and geometric shapes to help you visualize the placement of these stars. Work from a larger scale to a smaller scale as you do this to help maintain accurate proportions. For example, in **Figure 5.2.5**, using the circled stars as guides, I was able to plot the star marked A as the apex of a flat isosceles triangle. With that star marked, I plotted B using a linear relationship and C as the apex of a smaller isosceles triangle.

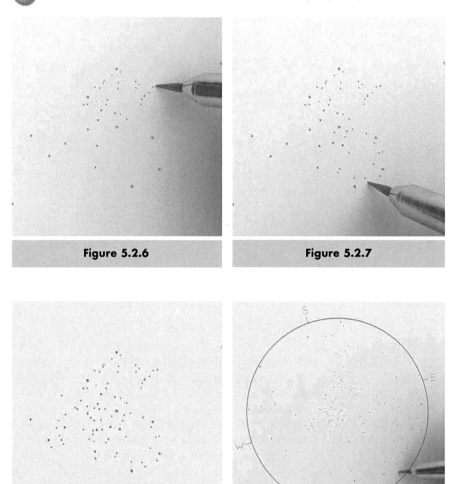

Figure 5.2.6

Figure 5.2.7

Figure 5.2.8

Figure 5.2.9

Continue this process around the main body of the cluster as demonstrated in **Figures 5.2.6** through **5.2.8**.

Step 4: Plotting remaining field stars
As with the core of the cluster, plotting the surrounding stars should be handled methodically. Use the same geometric relationships to place stars as you proceed clockwise or counterclockwise around the sketch. In this example, I began at the top and worked clockwise, as shown in **Figures 5.2.9** through **5.2.11**.

As methodical as you try to be, it is still important to recognize that you are a human being, not a camera, and that no sketch is going to be perfect. As long as you have taken care with your framework and used a consistent method of

Sketching Star Clusters

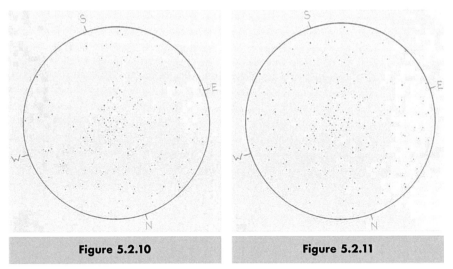

Figure 5.2.10   Figure 5.2.11

placing the rest of the stars, you will be able to produce a very good representation of the object you are sketching. Another factor to keep in mind is that you do not want to burn yourself out. This means that you may need to find a reasonable limit to how faint you are going to sketch in a very rich star field. The vicinity of M50 is drenched with faint stars, and although I could see many more stars, trying to sketch them all would not have made for a very enjoyable observation.

Step 5: Shading the base of unresolved starlight As you prepare to sketch this portion of the cluster, it is important that your eyes are as dark adapted as possible. Review *Sketching Faint Objects in Low Light* on page 42 for tips. Take your time observing the cluster away from your red sketching light. Concentrate on determining how this nebulous backdrop of faint starlight relates to the brighter members of the cluster. Is it a uniformly circular haze that drops off softly at the edges? Or is its shape uneven? In either case, which stars best mark its boundaries? Is it brighter toward the center? Or does it have clumpy patches of brightness? As you ask yourself these questions, work to burn the image, or a portion of the image, in your mind before dropping down to add it to your sketch.

For this view of M50, I perceived the soft glow to be roughly circular at the center of the cluster. However, rather than brightening toward the middle, it formed more of a U-shaped patch around the core. I decided to concentrate on the bottom edge of the soft feature and work my way up. Where you begin to lay down your shading will depend on what you are observing and your own preferences.

In order to add this nebulous patch of unresolved stars, scribble a dark palette of graphite somewhere outside of your sketch area with a B or HB pencil as discussed in *Using a Blending Stump* on page 153. Load your blending stump with graphite from this palette and lighten it a bit by scrubbing it in a blank area near the palette. With the stump ready, hold it at a shallow angle and, with a very light

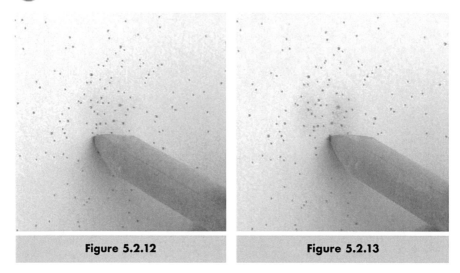

Figure 5.2.12   **Figure 5.2.13**

circular motion, begin to rub it against the sketch in the same area you observed the nebulous patch. (**Figure 5.2.12**) Take care not to press too firmly. Stronger patches of shading can be added a little later. For now, lay down a very light base of shading that defines the softer extents of the unresolved stars.

If the drop in brightness at the edges of this hazy region is very gradual, you will want to ease the pressure on your blending stump from very light to almost nothing as you make even, circular motions toward the edge. If the drop in brightness is more solidly defined, adjust your pressure accordingly.

> ### Which Comes First, the Shading or the Stars?
>
> When applying shading to a deep sky object, you have a choice to apply the shading before or after you have plotted stars in the area. I prefer to apply shading after many or all of the stars in the area have been plotted. Others prefer to insert the shading prior to filling the area with stars. There are benefits and drawbacks to both methods.
>
> If you add the shading after the stars have been plotted, you may have to contend with the shading process obliterating some of your fainter stars. It can also inadvertently pick up graphite from bolder stars and generate a dark halo around them. Both of these problems can be dealt with afterward by redrawing the fainter stars and using a kneaded eraser to lift out unwanted dark halos. The benefit of this method is that having most or all of the stars already plotted gives you a very precise framework on which to add your shading.
>
> If you generate the shading before filling in the stars, you avoid the problems noted above; however, you will not have a handy framework of stellar points on which to build. In this case, your shaded, nebulous patch will actually become the framework itself, to which the stars will adhere afterward. As always, there are many ways to approach a problem. Try both sequences and see what works best for you. You may find that you prefer one method over the other depending entirely on how the object appears through the eyepiece.

Sketching Star Clusters

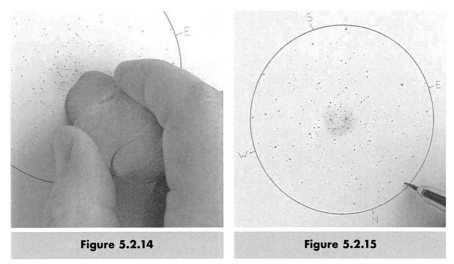

Figure 5.2.14   Figure 5.2.15

Step 6: Refining the unresolved starlight Once you have finished applying the overall boundaries of this shaded region, spend more time at the eyepiece to determine which areas, if any, are brighter than others. If the glow is very subtle and even, you may not need to add more shading. However, be sure to take your time to carefully observe this feature. If there is any unevenness to this misty light, it will greatly improve the quality of your sketch if you make an effort to reproduce it. If you do discern brighter patches, reload your blending stump, return to your sketch, and add a new layer of shading to the appropriate areas. Keep the pressure light and build it up slowly to keep the transitions soft. If you press too firmly, you may end up with hard-edged, unnatural blotches. Of course, while you are in the process of learning this technique, these hard edges may be difficult to avoid. However, it will get better with practice.

With the sketch of M50, I picked up not only the U-shape mentioned earlier but also a slightly brighter section along the eastern edge of the core. (**Figure 5.2.13**) Keep building up layers of shading as needed to present the full range of brightness in the cluster. Take some care that you do not make it too dark, since this may indicate a brighter patch of starlight than you actually saw. If it does get too dark, or if you get somewhat overzealous in the boundaries of your shading, it is time to break out the kneaded eraser. This is a fantastic tool for subtly removing shading. Mold the eraser into a shape that conforms closely to the area you want to erase and then lightly dab it down to lift off small amounts of graphite as described in *Using a Kneaded Eraser* on page 157. Continue doing this until you achieve the desired effect. (**Figure 5.2.14**) If you end up removing too much, you can easily come back in with the blending stump and reapply the right amount of shading. The kneaded eraser will not distress your paper and is perfect to use along with your blending stump.

Step 7: Finishing the sketch Once you have this shaded region where you want it, it may be necessary to replot some of your stars if they have been lightened or blurred in the process. (**Figure 5.2.15**) Finally, take another look

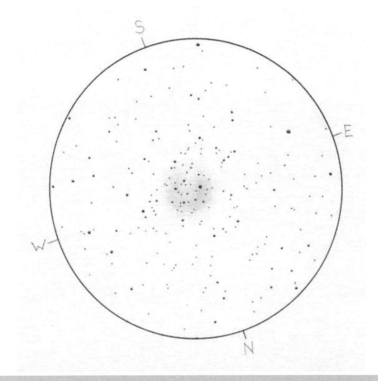

Figure 5.2.16

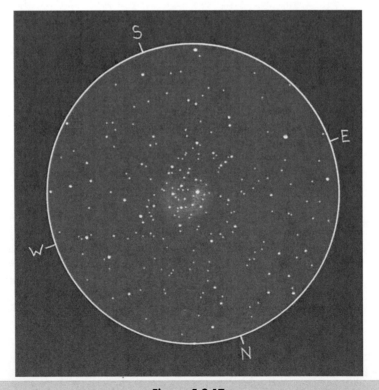

Figure 5.2.17

through the eyepiece and assess the brightness of the stars throughout the sketch. Distinguish them with levels of boldness that indicate their relative magnitudes. You may want to switch to an HB pencil at this point to mark the bolder stars. With that step complete, fill in any other important written details concerning your observation, and you are finished. (**Figure 5.2.16**) **Figure 5.2.17** shows the sketch inverted to better convey the view through the eyepiece.

Globular Clusters

On a clear summer night, from a dark location, you may decide to cast your unaided gaze on the constellation of Hercules. Along the western edge of the compact torso, or "Keystone" of this Greek hero, you might notice the soft fuzzy spot that marks the location of M13. Through binoculars, this outstanding globular cluster will appear more obvious as a round haze that is brighter at the center, but through a telescope it really shines. From a dark location, you can expect to be treated to a round puff of light that can take on the appearance of a mound of salt spilled on black velvet. Clumps and radiating arms of grainy starlight give this cluster a great deal of personality and daunting complexity for the astronomical sketcher.

Not all globular clusters are as large and richly detailed as M13, but they are all unique and beg to be explored through sketching. As we take a look at methods of rendering these wondrous objects, we can build on the skills used to sketch open clusters. Sketching a globular cluster, in some ways, is like taking the next step beyond an open cluster that contains a mist of unresolved stars. The tutorials presented here will rely on telescopic observations. If you are using binoculars, the same methods will still apply; however, the scale and detail of the clusters will simply be reduced.

There are a number of things to look for while observing a globular cluster prior to your sketch. Do you see any distinct stars across the face and edges of the cluster? Some clusters are noted for the clumped or stringlike patterns such stars make across their faces and fringes. Does the body of the cluster itself exhibit a smooth, mottled, or grainy appearance? Or perhaps it exhibits a combination of the three. How bright and condensed is the core of the cluster? Try to discern the overall shape of the cluster. At first glance, it may appear simply circular. However, after taking time to scan the view using averted vision, you may notice a subtle elongation or irregular extensions. How does the cluster relate to other stars in the vicinity? Make notes of these observations in a way that is convenient and useful to you, and then refer to them as you make the sketch.

5.3 Sketching a Simple Unresolved Globular Cluster

Let us begin by sketching a simple globular cluster that is not likely to exhibit graininess or a clumpy appearance through smaller-aperture telescopes. This tutorial will focus on a sketch of M75, which is one of the more remote Messier

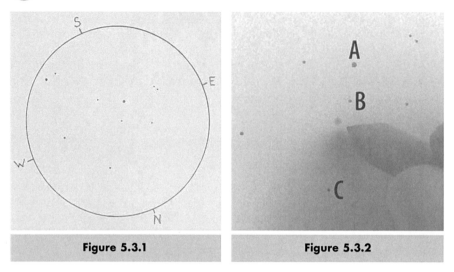

Figure 5.3.1   Figure 5.3.2

globular clusters. I observed this cluster with a 15-cm f/8 Newtonian, using a 10-mm Plössl eyepiece for a magnification of 120×. Through the 24-arc minute true field of view, there was no granularity or mottled structure to be seen. Its profile was soft and circular. The core was condensed, but not quite stellar in appearance, and appeared to have a somewhat harder edge along its southeast side.

Step 1: Framing and preparing the sketch area

If possible, frame the cluster so that an obvious star, or the cluster itself, rests at the center of the view. Next, mark the cardinal directions around the sketch area. Then lay in the framework of brighter field stars that are visible through the eyepiece. (**Figure 5.3.1**) Use the same methods discussed in the Open Cluster section for plotting these stars.

Step 2: Marking the position of the cluster

With the framework stars in place, it is time to pin down the location of the cluster within the sketch and shade it in place. There are a couple techniques that you can use to represent the cluster. You may choose to lightly shade it directly on to the paper with your pencil, blending it afterward with your finger or a blending stump. Or you may simply brush it directly onto the paper with a blending stump. I will describe the latter method in this tutorial.

To begin this part of the sketch, lightly load a blending stump with graphite as discussed in *Using a Blending Stump* on page 153. Then move to the sketch and softly brush in the position of the cluster's core. Use a light, circular motion to apply the shade. Apply just enough to lightly pin down its location and then compare it to the view through your eyepiece. If the position seems off, use a kneaded eraser to lift it off the paper and start over. This is the best time to make corrections in position. For this sketch, the core of M75 appeared to rest at the 7 o'clock position from star B and about a quarter of the way to star C in **Figure 5.3.2**. For globular clusters that do not have distinct cores, very lightly mark what appears to be the central hub of the cluster.

Sketching Star Clusters

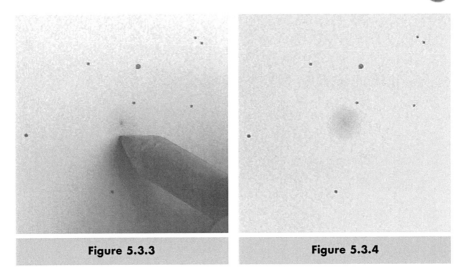

Figure 5.3.3 Figure 5.3.4

Step 3: Defining the boundary of the cluster Once you are satisfied with the position of the core, make an assessment of the cluster's diameter in the eyepiece. With smaller clusters, it can be easy to overdo the diameter if you do not have it planted firmly in mind. The framework of stars that you plotted comes in to play here. Look for convenient distances between those stars and use them to mentally describe the diameter of the cluster. If stars in the immediate vicinity of the cluster are of no use, look for others further out that you can use. For example, in this sketch, the cluster seemed to be just a bit narrower than the distance between stars A and B in **Figure 5.3.2**.

With your blending stump still very lightly loaded with graphite, use delicate, circular strokes to expand the size of the cluster to what you have determined from your observation. Work from the inside out. As you approach the outer rim of the cluster, lighten the pressure until you are just barely caressing the paper's surface. (**Figure 5.3.3**) This will give the cluster an appearance of fading very softly into nothing. If, however, the cluster appears to have a harder edge, adjust the pressure of your brush strokes accordingly.

Step 4: Building up bright areas Be sure to look for any irregularity in the structure of the cluster so you can represent that in the sketch. With M75, I noted that the outer reaches of its core were offset to the southeast and that this region appeared to be more sharply defined. With these thoughts in mind, move back to the sketch and add another layer of graphite. (**Figure 5.3.4**) Using slightly heavier strokes, swirl your shading from the center outward, lightening the pressure as you go. Avoid taking this layer of shading all the way to the edge of a previous layer if you want to maintain a soft transition. With this sketch, I kept the northern and western sections as soft as possible, but made the southeastern quadrant a bit more hard-edged.

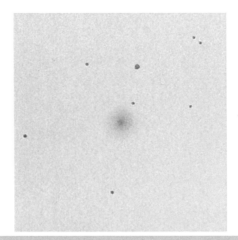

Figure 5.3.5

Step 5: Adding the core Return to the eyepiece now to determine how dark and concentrated you should render the core. For a loosely concentrated cluster, you may not need to add any additional shading. In fact, if the original spot you used to indicate the diffuse cluster's position still shows through your shaded layers, you may need to use your kneaded eraser to gently lift it away and leave a smooth surface behind. For clusters with some discernible degree of condensation, use your blending stump to gradually increase the shading at the core. Use loose, circular strokes to indicate a moderately condensed core. Use smaller, circular strokes and greater pressure to indicate one that is brighter or more strongly condensed. In either case, feather the dark core region softly into the surrounding halo to match the transition you see through the eyepiece. (**Figure 5.3.5**)

If the core is particularly bright and highly condensed, you may need to help it along by directly shading it with your pencil. Try an HB or B pencil and lightly swirl it in, then follow up with a circular application of your blending stump. Repeat this process if necessary until the core matches your eyepiece view. Although M75 was relatively condensed, the core was not bright enough to require direct pencil intervention. Even with the blending stump, I had to take care not to spike it too heavily. As you get used to the shading process, the application of these layers will become a more fluid process and will not always take place in distinct steps. You will move back and forth between the inside and outside of the cluster as you gradually build up the tone and approximate what you see through your telescope.

Step 6: Finishing the sketch Take another look through the eyepiece and assess the relative magnitudes of the stars you see. Then spend a few moments making any necessary adjustments to the stars in your sketch. Complete any notes about the observation, and with that finished, you can call your sketch complete. The final sketch of M75 is shown in **Figure 5.3.6**. An inverted, positive version can be seen in **Figure 5.3.7**.

Sketching Star Clusters

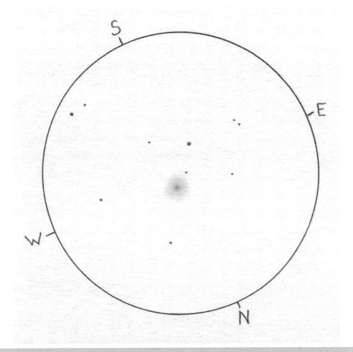

Figure 5.3.6

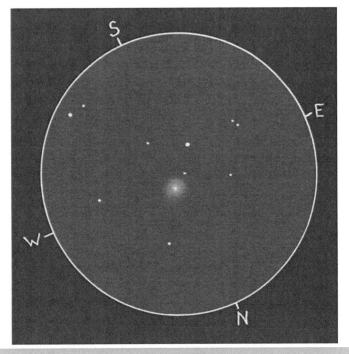

Figure 5.3.7

5.4 Sketching a Complex Globular Cluster

Not all globular clusters are simple patches of haze, of course. Quite a few are dazzling, visual delights that can be quite challenging to sketch. So let us break the process down into manageable steps and tackle one of these majestic clusters.

For this tutorial, we will consider M13, the Great Globular Cluster in Hercules. This bright, mottled, globular cluster is one of the best known in the Northern Hemisphere. When viewed at 120× through my 15-cm f/8 Newtonian under very dark skies, it was a delight to behold. It sparkled with flickering grains of starlight across its wide, mottled surface. Several thin streamers of stars appeared to flow away from its broad face, as though blown by a gentle cosmic wind. One moment it appeared as a luminous, potted fern with fronds draping toward the west. Another moment, it appeared as a scurrying spider.

With a globular cluster as detailed as M13, there are a variety of features that need to be observed and sketched. Although the steps listed below handle these features in a particular sequence, it is not my intention to imply that this is the only approach you should take. The order in which you sketch such features will vary depending on your sketching style and the actual traits of the cluster you are observing.

Step 1: Preparing the sketch and marking framework stars Sketching this cluster begins, as before, by marking cardinal directions around your sketch and framing the view to best display the cluster. With a large and complicated cluster such as this, it may be difficult to find a single star to place at the center. Although it is helpful to have a notable star there, it is not necessary. With M13, I was satisfied with placing the body of the cluster roughly in the center. In this situation, you do not have a convenient central anchor to help visualize the imaginary clock face, and concentric circles that help place framework stars. However, with some concentration, you can still visualize these structural aids. With the framing of the view and marking of cardinal directions taken care of, you can plot the stars seen in the eyepiece. They will serve as the latticework upon which the cluster finally rests. (**Figure 5.4.1**)

Step 2: Marking distinct stars within the cluster Depending on your observing conditions, your optics, and the cluster itself, you may be able to pick out a few, or even numerous, distinct stars across its surface. Taking time to study these stars and their positions relative to each other will provide you with a better appreciation of the cluster's unique character. Plotting a reasonable number of them in your sketch will help to record that character. If you are able to resolve huge numbers of stars here, it will be impractical to try and plot them all. So if this is the case, choose several of the brighter members and just record them. Use the techniques described in the open cluster tutorials to position these stars by relying on geometric relationships and relative distances compared to other stars that you have already plotted. With this sketch of M13, I plotted 27 stars that I was able to resolve distinctly above the grainy glow of the cluster itself. (**Figure 5.4.2**)

Sketching Star Clusters

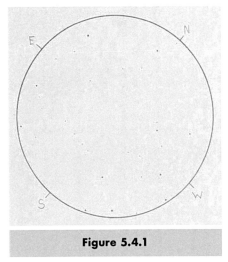

Figure 5.4.1

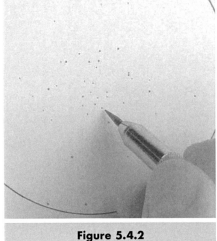

Figure 5.4.2

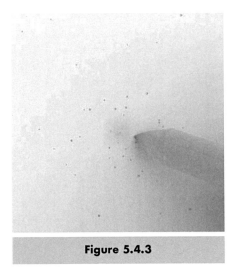

Figure 5.4.3

Step 3: Examining the boundaries and position of the cluster In the next few steps, we will concentrate on suggesting the unresolved, grainy glow of the cluster by using a blending stump loaded with graphite. Using the stars that you have previously plotted as a guide, examine the core and the halo of the cluster. Use averted vision to get the best impression of where the outermost boundaries of the cluster rest in relation to these framework stars. You will also want to note where the core of the cluster lies in relation to these stars and examine the position of any irregular clumps of brightness. Do not worry about trying to remember all of this at once. In the following steps, we will approach these individual pieces one at a time.

Step 4: Marking the core If the cluster exhibits a bright core, this is one place where you can choose to begin shading your sketch. Load your blending stump with graphite and apply it to the sketch with a smooth, circular motion. Compare its position carefully to the surrounding stars so that it can serve as an accurate anchor for the rest of the shading process. Keep the outer edges of this core region soft for now so that it will transition smoothly into the other features that will be added next. **(Figure 5.4.3)** If the cluster you are observing is very diffuse and does not offer a distinct core, you can skip over to the next step.

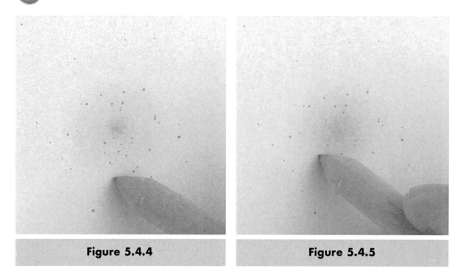

Figure 5.4.4 Figure 5.4.5

Step 5: Establishing the halo Now, using the framework stars as a guide, shade in the full reach of the cluster's luminous halo. Work from the inside out, keeping the pressure light and getting lighter as you approach the outer edges. Try to keep these outer boundaries soft as they fade into nothing. (**Figure 5.4.4**) Continue to apply shading from the inside out to establish the basic brightness profile that the cluster exhibits. You do not have to be very detailed at this point; simply apply gradual shading to indicate where the cluster is brightest. (**Figure 5.4.5**)

Step 6: Indicating luminous filaments If the cluster you are observing exhibits radiating strings of stars, you can indicate them with a light touch from your blending stump. Take care to note how they relate to the brighter stars that you plotted at and around the face of the cluster. Be sure your blending stump is very lightly loaded with graphite before you do this. Gently apply these features with curved or linear strokes that follow the shape of the strings. As much as you try to avoid it, it can still be very easy to overdarken these delicate features. If that happens, you can use a kneaded eraser to gradually lift away the excess and leave behind the subtle stream of stars you want. During my observation of M13, I perceived seven of these gently arcing strings, as shown in **Figures 5.4.6** through **5.4.8**.

Step 7: Adding clumps of brightness If you detect irregular areas of brightness during your observation, note how they relate to the overall body of the cluster and any stars you plotted within. Then load your blending stump with more graphite and gently apply these features. Where the edges of these regions are soft, ease up on your blending stump's pressure to maintain a gradual transition in shading. The sequence in which you render these clumps is up to you. With M13, I began with the large, bright patch at its heart and moved out from there, as shown in **Figures 5.4.9** through **5.4.11**.

Step 8: Refining overall brightness As you examine the view through the eyepiece and compare it to your sketch, see if you need to refine the amount

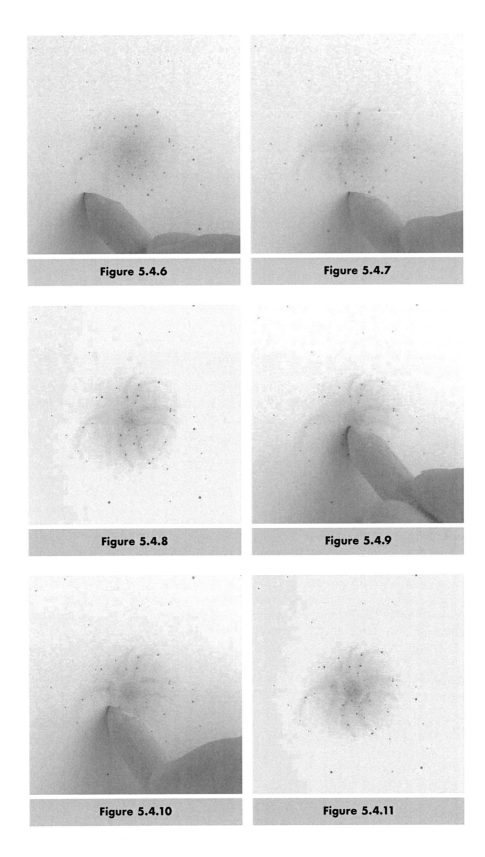

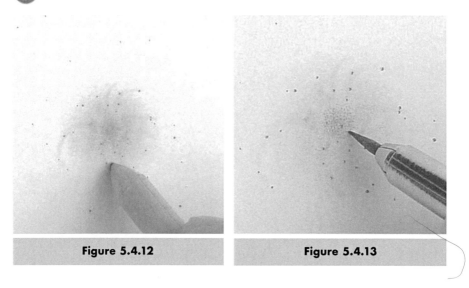

Figure 5.4.12   Figure 5.4.13

of shading that you have applied so far. If some of your transitions are too blunt, or if the cluster needs to be darkened or lightened in certain areas, take care of that now. Use your blending stump or kneaded eraser to accomplish these tasks. With this sketch, I used my blending stump to darken and expand the outer reaches of the halo. (**Figure 5.4.12**)

Step 9: Indicating granularity in the core If the cluster you are observing sparkles with dozens or hundreds of tantalizingly faint stars, you can take steps to add that effect to your sketch. To represent the teeming masses of flickering starlight, you can use a careful stippling technique. Stippling is a method of randomly applying dots to a drawing to build form and texture. Caution is required with this technique though, since stippling can easily give an unnecessarily artificial look to your sketch. It may also give the impression that those random dots are real, observed, individual stars. For these reasons, some observers prefer not to take this step; they detail the granularity in their written notes instead. With practice however, stippling can add a degree of depth and sparkle to a globular cluster sketch that better represents what you saw through your telescope. For tips on practicing this useful technique, see *Stippling Technique* on page 129. If the cluster does not show any granularity, or if you would rather not use this technique on your sketch, you can proceed to the last step in the tutorial.

Before you begin rendering these granular areas, take a look at the stars that you originally plotted within the cluster. As you went through the blending process, some of them may have been blurred or faded. It is important that these distinctly observed stars stand out from the granular field that you will add next. So take a few moments now to darken them back up with your pencil.

I suggest using 4H pencil when you add stippling. The harder lead will give you lighter dots that this subtle effect begs for. Begin by stippling into the brighter regions of the cluster. The spacing of the stippled texture should be tightest in these areas. (**Figure 5.4.13** and **5.4.14**) After working on the brighter areas, apply

Sketching Star Clusters

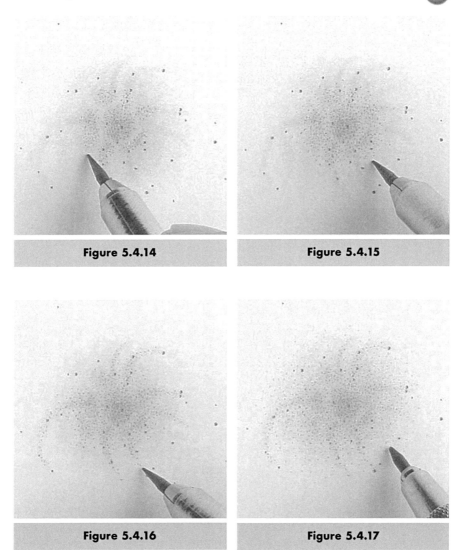

Figure 5.4.14

Figure 5.4.15

Figure 5.4.16

Figure 5.4.17

stippling to the spaces between and around the outside of these clumps. Use a wider spacing between the dots to indicate a slightly fainter appearance. (**Figure 5.4.15**)

Step 10: Indicating granularity in the periphery
If you used shading to hint at strings of stars emanating from the cluster, you can judiciously add stippling to these as well. Start off conservatively as you add lightly stippled dots to them, since they can easily begin to look much more solid than you intend. (**Figure 5.4.16**)

Next, add stippling to the outer reaches of the halo. As you move away from the brighter regions of the cluster, the spacing of your dots should get wider and wider to achieve the effect of the granularity fading away at the edges. (**Figure 5.4.17**)

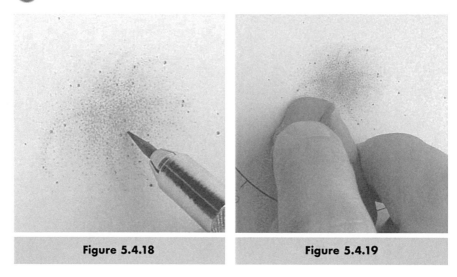

Figure 5.4.18   Figure 5.4.19

Step 11: Refining the stippling As you examine your sketch and compare it to your telescopic view, see if the overall brightness of the stippling is true to the values seen through the eyepiece. At this point, you may find it helpful to darken or increase the concentration of the stippling in the brighter areas of the cluster. To accomplish this, you can switch to a 2H pencil to get a slightly darker dot. You can now begin to reapply stippling to any areas that need it. Be sure to arrange the dots around the edges of these areas with gradually wider spacing so that the transitions are not too blunt. (**Figure 5.4.18**) A potential trouble spot when you do this is that some of the specific stars that you plotted earlier may start to get a bit lost. If you see that happening, pause for a moment to darken those stars up just a bit. This will ensure that they continue to appear distinct from the stippled masses around them.

If you find that the stippling has been applied too heavily in places, use your kneaded eraser to gently lift away the graphite until it looks right. (**Figure 5.4.19**)

STEP 12: Finishing the sketch Return to your eyepiece and compare the brightness of the individual stars that you plotted earlier to what is in your sketch. Darken them where necessary to match what you see through your telescope. Then finish writing notes about the observation, and your globular cluster sketch is complete. (**Figure 5.4.20**) A positive version of this sketch can be seen in **Figure 5.4.21**.

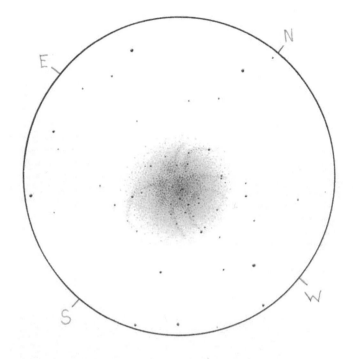

Figure 5.4.20

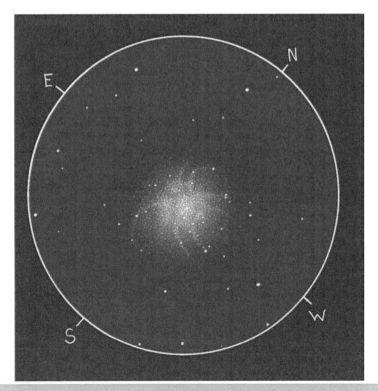

Figure 5.4.21

Tips and Techniques

5.5 Marking Stars

Just as the varying brightness of stars adds structure, beauty, and depth to your view of an astronomical object, so will the care you give star weights in your sketch. Keep in mind as you sketch that the brighter a star is, the bolder you should render it. To begin with, you may decide to divide your assessment of star magnitudes into three categories: faint, medium, and bright. The faint stars can be created with a light press of the pencil. The medium brightness stars can be made by pressing the pencil more firmly and rotating or twirling it slightly. Bright stars can be created by pressing the pencil firmly, carefully circling it around that point, and then gradually widening the circle into a bold dot. Take care to keep the larger dots as round as possible.

Some star fields may contain a very wide range of star brightness and require a greater variety of star weights. To handle these situations, you may want to use a pencil with a harder lead, such as H or 2H for the faintest stars. Then switch to a softer lead such as an HB or B to mark the mid-range to brighter stars. When you begin your sketch, try to keep the weight of your stars conservative and then bulk them up as needed. It is easier to make a star heavier than it is to erase it and redraw it lighter.

Some observers adopt a very systematic approach to assessing star brightness, carefully comparing stars to one another and then maintaining this consistency very carefully as they develop the sketch at the eyepiece. This method can be invaluable if you want to convey the magnitudes of variable stars, supernovae, or any other object that you want to estimate with your own eyes. Other observers like to refine their sketches later by checking their rendering against a star atlas, noting star magnitudes and applying those weights to the sketch as a final touch.

You may find it helpful to first practice drawing stars in the brightly lit comfort of your home. Try creating rows of stars in different weights on a sheet of paper. Your goal should be to get comfortable generating point like or circular dots. That way, when you are in the dark, with only a faint light to guide your hand, you are not struggling as much with the mechanics of plotting a star as you are with studying the object and transferring its geometry and aesthetics to paper. If you prefer to hold your pencil in a normal "handwriting" position (**Figure 5.5.1**), take care to keep the pencil from flicking as you lift it up from the paper. It is easy to get streaks instead of dots holding the pencil this way. Others prefer to hold the pencil vertically to the page, press it down, and twirl it slightly. (**Figure 5.5.2**) This method makes distorting the shape of the stars less likely and can be a very good place to start. Try both methods as you practice and see which works best for you. If you do hold your pencil at an angle, try to rotate it an eighth to a quarter turn every few stars so that the point gets worn down evenly.

The sharpness of your pencil is an important part of this process. Whether you are using wood pencils or lead holders, you will want to have a sharpener on

Sketching Star Clusters

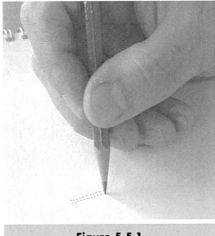

Figure 5.5.1

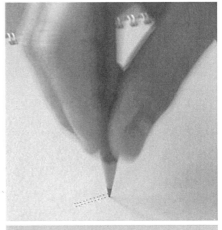

Figure 5.5.2

Figure 5.5.3

hand. A dull pencil or one with an odd-shaped tip will make it difficult to accurately plot stars. On the other hand, an overly sharp pencil can be difficult to control and the tip may break and mar your sketch. After sharpening your pencil, you can scribble it around in a blank area outside of your sketch to take off a bit of the edge. Another useful tool to have on hand is a sanding pad. Gently brush your pencil across the face of the sanding pad, steadily turning the pencil as you go, to evenly hone the tip to a manageable point. **(Figure 5.5.3)**

5.6 Correcting Misplotted Stars

As hard as you try, there are going to be times when you mark a star in the wrong spot and need to correct it. How easy a correction it is will depend on the type of paper and pencil you are using, as well as how bold you made the star. If you find it difficult to correct mistakes at the telescope, you may choose to mark the star for deletion later by lightly putting a slash or an X through it. Whether you correct the mistake at the telescope or indoors, the first thing you should reach for is your eraser and, if needed, your eraser shield. If the star is sitting all by itself, you can erase it directly. However, if it is in close proximity to another star, your eraser shield will come in handy. Find a small hole in the shield and place it over the star so that other nearby stars are protected. All you have to do now is to start erasing over that hole. (**Figure 5.6.1**) You may need to lift the shield up a few times and reposition it to ensure that the errant star is erased as completely as possible. After erasing the star, avoid using your hand to remove any eraser crumbs, since this can smear your sketch. Blowing on the sketch also risks getting moisture on it. Instead, keep a brush handy and use it to remove the debris.

Sometimes a light after-image of the star will remain after erasing. If you would like to delete it further, you may find success using an exacto knife to lightly scrape away the upper layer of paper. Hold the knife at an angle and *very* lightly scrape the tip against the faint remnants of the star. (**Figure 5.6.2**) Be careful not to dig too deeply or the fix will look worse than the problem. Just gradually work the knife tip side to side across the star until the pencil mark has disappeared to your satisfaction. Some papers will take to this better than

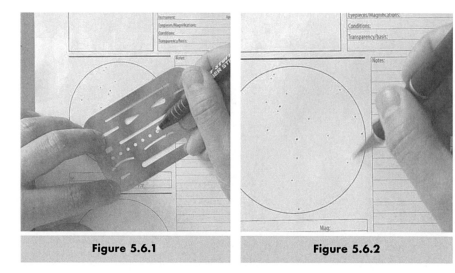

Figure 5.6.1   **Figure 5.6.2**

Sketching Star Clusters

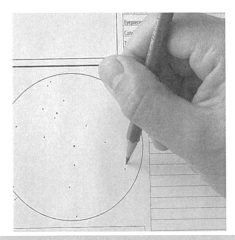

Figure 5.6.3

others. It is a good idea to practice this ahead of time so that you get a feel for it and know how your paper responds. Finally, replot the star where it should be. (**Figure 5.6.3**)

Now, two words of caution:

- If you anticipate that you will later add shading across this corrected portion of your sketch, take care how strongly you erase the star, and definitely do not use the knife technique. Any roughening of the paper will cause that portion of the sketch to appear unnaturally darker when you shade it later. If there is still a light after-image of the star, the shading you add may very well hide it.
- If you experience heavy dew during your sketch, be careful, since erasing damp paper can destroy the paper or simply smear the star. In situations such as this, it may be better to mark the star and take care of erasing it indoors the next day, after the paper has had a chance to dry out.

5.7 Stippling Technique

Stippling is a helpful technique to indicate granular star fields, especially with globular clusters. Before heading out into the night though, take some time to practice stippling so that the process becomes more natural at the eyepiece. Feel free to use any weight of pencil for practice, although for final sketches you will likely want to use hard 2H to 4H leads that produce light, subtle dots.

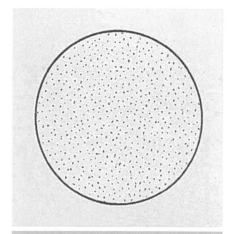

Figure 5.7.1 A semirandom stipple pattern

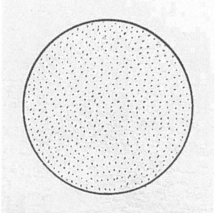

Figure 5.7.2 An overly structured stipple pattern

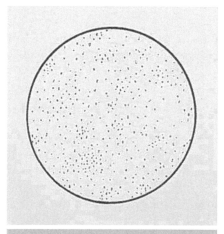

Figure 5.7.3 An uneven stipple pattern

Figure 5.7.4 A dense, but lightly applied stipple pattern

The key to successful DSO stippling is the ability to generate a semirandom spread of dots. **(Figure 5.7.1)** If the dots develop a pattern of rows or arcs, they can appear artificial and distracting. **(Figure 5.7.2)** On the other hand, if they are spread too unevenly, they can give the unintentional appearance of mottled structure. **(Figure 5.7.3)** If you have not tried stippling before, achieving this semirandom effect can be challenging. However, with patience and repeated practice, you will get the hang of it.

Sketching Star Clusters

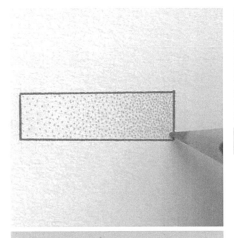

Figure 5.7.6 Stippled practice clusters

Figure 5.7.5 A graduated range of stippling

Your first goal should be to get comfortable applying the dots in a uniform, but unordered fashion. Try to fill a small circle or square evenly with stippled dots. Instead of working methodically from one end to the other, work loosely. Add a few dots widely spaced off to one side, then freely add a few more dots to fill the loose gaps in between. Keep the pattern fairly open. Try not to overthink it. Work quickly with the goal of becoming almost reflexive. Try slightly blurring your vision occasionally to help see where the pattern is uneven.

After practicing that a few times, try again with more tightly spaced dots. (**Figure** 5.7.4) For the typical sizes of deep sky sketches, stippling will involve fields of dots that are fairly dense, even though they are lightly applied.

As you progress, try filling a rectangular area with stippled dots that are dense at one end and gradually become more widely spaced across the length of the rectangle. (**Figure** 5.7.5) Once you get a feel for this, you will be able to suggest varying densities of stellar granularity. Try practicing imaginary globular clusters by stippling heavily at the center of a circle and widening the stipple pattern as you move to the outside of the circle. (**Figure** 5.7.6)

CHAPTER SIX

Sketching Nebulae

Few things attract newcomers to visual astronomy and set them up for frustration like the crisp, vivid photographs of galactic and planetary nebulae that amaze us and fuel our imagination. The actual views through a telescope are, of course, far more subtle, requiring patience and persistence to observe and appreciate. Although the details in these views can often rest on the edge of detection, with practice they can be very rewarding subjects, since you are perceiving those gargantuan swirls, streamers, and plumes of interstellar matter with your own eyes.

To increase your ability to view these objects and the structures they contain, you will want to use every observing technique at your disposal. Nebulae are easily washed out by light pollution, so finding a dark location from which you can observe will greatly assist the amount of detail that you can see. Make sure to allow time for your eyes to become dark-adapted. Use averted vision to detect the subtlest variations in brightness. If you have funds available, investing in quality eyepieces and nebula filters will help you expand the level of detail you can discern. Beware that even if you use the tips stated above, there is much you will miss if you do not take your time with an observation. This is where sketching becomes a fantastic means of focusing your attention.

Recording these sights in a sketch can add an extra layer of difficulty to the observation, but it is well worth the effort. Whether you are trying to capture the sublime beauty of a showpiece nebula or testing the limits of your observing skills, sketching what you observe will help you achieve these goals. In this chapter, we will discuss sketching methods for diffuse nebulae, planetary nebulae, and dark nebulae.

The sketching materials you choose can be as basic as a pencil, paper, clipboard, and red light. However, to help refine your sketch, there are a few additional materials you may want to include. Following is a suggested list of materials for producing a graphite nebula sketch:

- Clipboard
- Dimmable red observing light
- Paper prepared in any of the following ways:
  - Blank
  - Prepared with predrawn sketching circles
  - Copied or preprinted log sheets
  - Copied, printed or traced star fields (as discussed in the tutorials on pages 32 and 145)
- HB and 2H pencils
- Pen (for notes)
- Blending stump or tortillon
- Choice of erasers (Art Gum, eraser pencil, kneaded)
- Eraser shield (used to constrain erasures to a small area)
- Pencil sharpener or lead pointer
- Sandpaper block (used to hone the point of a pencil, blending stump, or tortillon)
- Small paint brush (used to brush away loose graphite or eraser debris)

6.1 Sketching a Diffuse Nebula

Diffuse nebulae, including supernova remnants, present delicate, often complex structures through a telescope. Many can be seen through binoculars and even a few stand up to naked-eye scrutiny. If you have noticed a fuzzy patch of light in the sword of the constellation Orion, you have made a naked-eye observation of M42, the Great Orion Nebula. The Lagoon Nebula, M8, in the constellation Sagittarius is another diffuse nebula that you can spot without optical aid under dark skies. Through binoculars, these brightest nebulae begin to come to life as glowing blossoms of mist, often peppered with blazing stars. Through a telescope, they can be spectacular. The wide variety of nebulae accessible to a telescope only get fainter from here, but not any less interesting. Like snowflakes and fingerprints, each one is unique. Because these objects assume such a wide variety of forms, there is no real cookie-cutter way to approach a sketch. Instead, we will cover broad techniques that you can apply to any nebula that graces your eyepiece.

As you observe a nebula, here are some things to keep in mind. Try not to assume that you already know the shape and details of the object. Soak in the view with the thought that you might see something new or unexpected. You might be surprised by what reveals itself when you focus your attention and keep an open mind. Averted vision and slight motions of the telescope can help you detect very faint variations in contrast. The entire nebulous region may not be continuous, so keep your eyes ready to detect patches of nebulosity that are detached from the main body. As the full extent of the nebula becomes more apparent and its finer details emerge, ask yourself which portions are brighter. How do they interact with the stars surrounding and overlapping them? How are their contours defined? Are they soft in some places and sharp in others? How does the nebula react to different eyepiece and filter combinations? With these things in mind, you will be better prepared to begin the sketching process.

Sketching a Diffuse Nebula

For this tutorial, we will be focusing on a combination emission/reflection nebula known as M20, or the Trifid Nebula. At first glance through my 15-cm f/8 Newtonian with a 10-mm Plössl eyepiece, it appears as a soft, somewhat irregular patch of fog. Further scrutiny of the 120×, 24-arc minute true field of view reveals a two-lobed structure aligned north-south. Each lobe is roughly circular with irregular patches of dark intrusions. The northern lobe blossoms away from a brilliant yellow-orange star, whereas the southern lobe is centered on a bright double-star. Applying an ultrahigh contrast filter to the eyepiece does some interesting things to the nebula. The northern lobe is almost entirely eliminated from sight, with only the barest hint of haze around the bright star at its heart. The southern lobe, on the other hand, begins to reveal finer structure. The dark intrusions hinted at without the filter now appear to cross the entire face of the nebula from three main directions. Continued observation shows the western intrusion to have a branch at the outer edge.

Although we cover dark nebulae more specifically later in the chapter, it is interesting to note that these intersecting dark rivers are actually a dark nebula in the foreground known as Barnard 85.

Step 1: Framing and preparing the sketch area Before you start your sketch, begin by centering your view in a way that best frames the nebula. If some slight adjustments will allow you to place an obvious star at the center of the field, try to do so, since this can be very helpful when plotting the remaining stars. Begin the sketching process by marking the cardinal directions around your sketch circle (see *Assessing Cardinal Directions* on page 39).

Step 2: Plotting the star framework The next step is to plot the positions of stars in the view. Tutorials for rendering star fields can be found in the *Sketching Star Clusters* chapter. Although the star field may not seem to be the most compelling part of the view, it is still an important part of the sketch. Not only does it lend context, but it also provides a framework across which you will stretch the misty shades of the nebula. It is important to be as accurate as

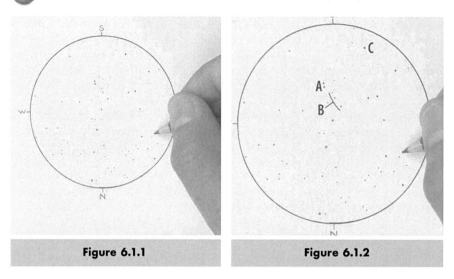

Figure 6.1.1

Figure 6.1.2

possible when plotting these stars, since they will help you define the nebula's size and shape. (**Figure 6.1.1**)

Step 3: Analyzing the nebula With expansive diffuse nebulae, deciding where to begin sketching can be a tough choice. There is not always a single core region to center upon. Since one of the important aspects of nebula sketching is to render the dimensions of the object accurately, you may want to start with a portion of the nebula that has the most obvious relationship to the stars you have plotted.

After adding my UHC filter to improve contrast in the emission portion of the nebula, I analyzed the southern lobe of M20. I noticed that its southeastern wedge aligned along some key stars. Its western tip just overlapped the central double-star marked A in **Figure 6.1.2**. Its northwestern edge ran along a row of stars marked B. Its southwestern edge ran from the double-star toward another star marked C and faded out about three-quarters of the way there.

Step 4: Brushing in the base layer Using the techniques described in *Using a Blending Stump* on page 153, lightly load your blending stump with graphite. Using a few key framework stars as a guide, take the stump and lightly swirl shading into the area that the nebula occupies. The best way to approach this is with a layering process. Start with a very light layer of shading to define the boundaries of one continuous section of the nebula. Use small circular motions to lightly push the graphite up to the boundaries that you perceive in the eyepiece. For boundaries that fade very gradually to nothing, gradually lighten the pressure on the blending stump to almost nothing as you approach that edge of the nebula. For this slice of M20, the northwestern and southwestern edges had relatively sharp edges, but the southeastern side faded away gradually. (**Figure 6.1.3**)

Step 5: Building up brighter areas Once the initial layer of shading is in place, build up areas of the nebula that need to be shaded more heavily. Add

Sketching Nebulae

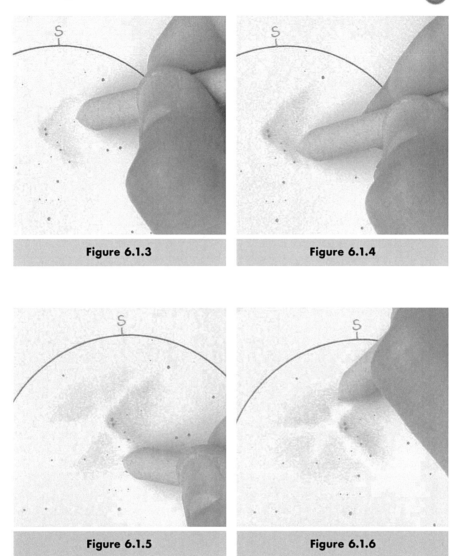

Figure 6.1.3

Figure 6.1.4

Figure 6.1.5

Figure 6.1.6

more graphite to your blending stump when necessary. This can be seen in the M20 sketch, where I came back in with more graphite and darkened the western and especially northwestern portions of the nebulous wedge. (**Figure 6.1.4**)

Step 6: Continue building up brighter areas Continue to blend in different sections of the nebula, using your framework of stars to guide the contours of light and dark. (**Figures 6.1.5** through **6.1.7**) You may choose to build up layers completely in each section of the nebula as you work on it; or you can lay down the initial light boundary layer for the entire object before going back and building up progressively darker layers all around. The order in which you

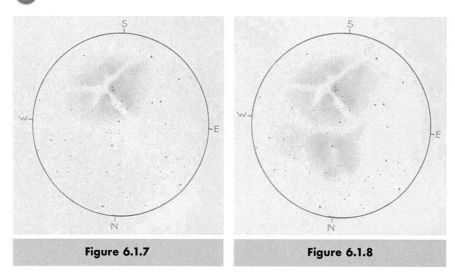

Figure 6.1.7   **Figure 6.1.8**

do this may vary from object to object or may simply be a matter of personal preference.

Step 7: Refining the view With the southern section complete, I removed the UHC filter to better view the northern reflection nebula. Following the same process, I built up layers of graphite to define its shape and varying luminosity. (**Figure 6.1.8**) As you compare the view through the eyepiece to what you see on the sketch, you may find areas that were shaded too heavily. You may also want to clarify or soften edges of the nebula. A kneaded eraser will allow you to subtly remove layers of graphite by gently dabbing or brushing it against the sketch, as discussed in *Using a Kneaded Eraser* on page 157.

Step 8: Rejuvenating stars During the shading process, a number of stars will probably be harmed. So get your pencil and rejuvenate them. This is also the time to reexamine the view through the eyepiece and assess relative star magnitudes. Return to your sketch and darken stars as necessary to represent these differences in brightness. (**Figure 6.1.9**)

The differing views of M20 offered with and without the UHC filter raise an interesting point. If you have several deep sky filters and take the time to examine an object using each one, how can you represent any differences in a sketch? There are a few possibilities. You may decide to sketch the object using the view that best represents the object. You can also create a *composite* sketch that combines the most detailed features visible in each view, or you can create *multiple* sketches detailing how the object appears in each view. The choice, of course, is completely up to you. Just be sure to make notes describing how you proceeded so that you can recall how you arrived at the sketch later. In this instance, I initially combined the best of both the filtered and unfiltered views. However, later, I digitally edited a copy of the sketch to show the effect that the UHC filter had on the entire object. Both finished sketches can be seen in **Figures 6.1.10** and

Sketching Nebulae

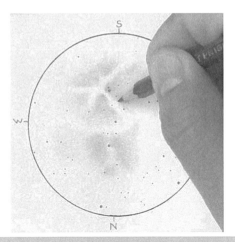

Figure 6.1.9

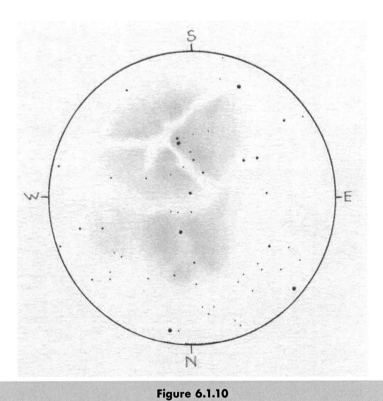

Figure 6.1.10

Astronomical Sketching: A Step-by-Step Introduction

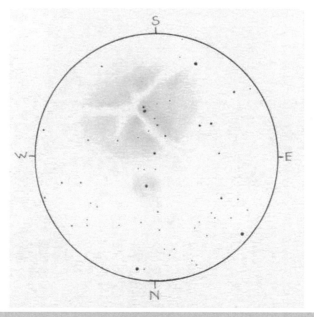

Figure 6.1.11

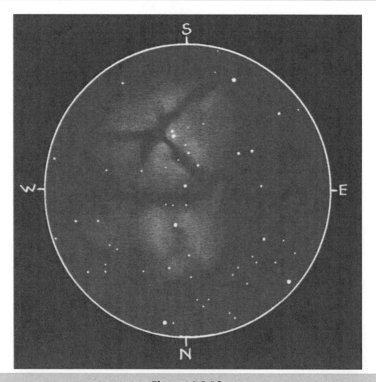

Figure 6.1.12

Sketching Nebulae

6.1.11. A positive white-on-black version of the composite sketch can be seen in Figure 6.1.12.

6.2 Sketching a Planetary Nebula

Compared to the diffuse nebulae we discussed in the previous tutorial, planetary nebulae tend to be compact, symmetrical objects. One famous example of a planetary nebula is the bright, hollow oval of the Ring Nebula (M57). Another is the large, circular form of the Helix Nebula (NGC 7293). Planetary nebulae tend to be circular or elliptical in form, although this is not always the case. However, do not be lulled into thinking that a simple, diffuse, disk of mist is all you will be sketching when you observe one of these objects. Within many of these gaseous globes, fascinating structures can be found. Others are no more than starlike objects, distinguishable from their true stellar neighbors through the use of a UHC or OIII filter that dims the surrounding stars, but leaves the nebula relatively untouched.

As you observe a planetary nebula, here are a few things to look for. How large and bright is it? What is its shape, and is it smooth or irregular? How is its brightness distributed? Are there any structures or mottling visible within it? Does it exhibit a ringlike appearance? Can you discern its central star, and if so, how bright is it? A number of planetary nebulae are bright for their small size, and you may be able to discern some color; so keep your eyes open for this. These objects also tend to respond well to high magnification, so be sure to observe with a high-power eyepiece to see if you can pick up details that might be missed at low power.

Sketching a Planetary Nebula

This tutorial will feature a sketch a beautiful planetary nebula, NGC 2392, in the constellation Gemini. Sometimes known as the Eskimo Nebula or Clown Face Nebula, it conveys a double-shell structure that tantalizes with mottling in the brighter inner shell. Like many planetary nebulae, it stands up well to higher magnification. I used a 10-mm Plössl and 2× Barlow for a magnification of 240× and a true field of view of 12 arc minutes. A bright star blazed a little more than one arc minute north of the nebula and helped me estimate its dimensions.

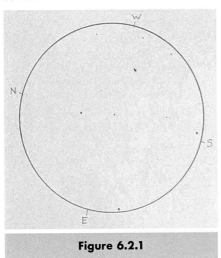

Figure 6.2.1

Step 1: Preparing the view and plotting stars To begin this sketch, mark the cardinal directions and then plot the framework of stars that fill the view. (**Figure 6.2.1**) Refer to

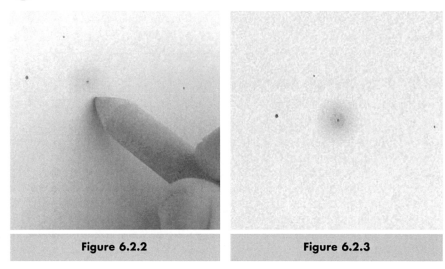

Figure 6.2.2 Figure 6.2.3

the *Sketching Star Clusters* chapter on page 97 for an in-depth discussion of this process. In the case of this planetary nebula, one of those stars was the central star of the nebula itself. I chose to place this at the center of the view. However, if you find a nearby object to be compelling, you may want to leave the planetary nebula off-center in order to include this object in the view.

Step 2: Examining the nebula Take a long look in the eyepiece to acquaint yourself with the dimensions and features of the object. If it is somewhat small and located in a sparse star field, it may be difficult to determine its size relative to stars that you have plotted. It may be necessary to look at closely spaced stars and estimate how the diameter of the nebula relates to the distance between those stars. In the case of NGC 2392, its radius appeared to be between a quarter and a third of the distance between the nebula's central star and the bright star just to the north.

Step 3: Defining the boundary As with the diffuse nebulae in this chapter, I like to use a blending stump loaded with graphite to brush in the nebula. See *Using a Blending Stump* on page 153 for tips on using this sketching tool. Because very fine details may be visible, try to use a blending stump that is well sharpened.

With the image from Step 2 well in mind, load your blending stump lightly with graphite and return to your sketch. Softly swirl in shading to produce an initial layer that defines the shape and size of the nebula. (**Figure 6.2.2**) Be mindful of how soft or sharp the outer boundaries are. If they are soft, be sure

Sketching Nebulae

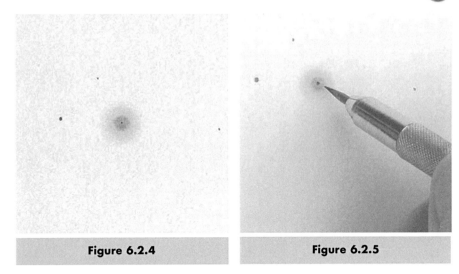

Figure 6.2.4   Figure 6.2.5

to keep pressure on the blending stump very light and dwindle it to nothing as you come to the edge of the nebula.

Step 4: Building up brighter areas Return to the eyepiece and reacquaint yourself with where any brighter spots of the nebula are. Use your blending stump to "brush in" those heavier shades. For simple nebulae, this can be a fairly quick process. Large, complex objects like M27 or NGC 7293 (the Helix Nebula) may need to be treated more like a diffuse nebula. For NGC 2392, the brighter, inner shell of the nebula suggested mottling and was elongated ever so slightly north to south. (**Figure 6.2.3**)

Step 5: Refining the nebula Continue to add overlapping layers of shading to build up the brighter portions of the nebula. Take care not to enlarge the size of these areas as you add more graphite. (**Figure 6.2.4**) You may also want to make use of a kneaded eraser to convey lighter details or to make corrections to shading that is too dark or in the wrong place.

Step 6: Replotting distressed stars With the shading complete, check any stars that are overlapped by the nebula. If they were lightened or blurred by the shading process, replot them with your pencil. (**Figure 6.2.5**)

Step 7: Finishing the sketch Finish the sketch by darkening stars as necessary to match brightness levels you see through the eyepiece. Add any notes

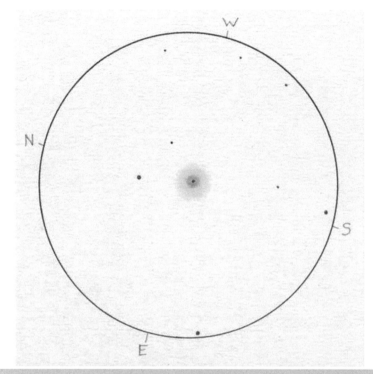

Figure 6.2.6

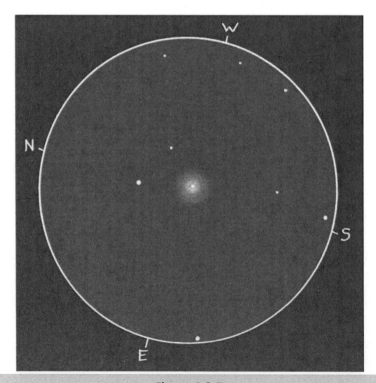

Figure 6.2.7

about the observation and your sketch is complete. (**Figure 6.2.6**) A positive white-on-black version can bee seen in **Figure 6.2.7**.

Dark Nebulae

Although dark nebulae do not receive a great deal of attention, they are everywhere. These opaque clouds of dust and gas require a luminous backdrop to be revealed to our eyes. In some cases, contrast around the dark nebula is so subtle that visual observation is exceedingly difficult, requiring very dark skies and a keen eye to discern. However, not all dark nebulae are so difficult to see. In some cases, they are silhouetted by an object that is so engaging itself that the dark nebula may be ignored. Ironically it is often this same dark nebula that provides the compelling contours that make the background nebula or star field appear so interesting. From the tiny rivers that cleave into the Trifid Nebula discussed earlier in this chapter to the vast currents of dark that cut through a naked-eye view of the Milky Way, dark nebulae are definitely worthy of our scrutiny and sketches.

Since a dark nebula represents an absence of starlight or nebulosity, figuring out how to sketch one may seem like a daunting task. However, if you have picked up techniques for sketching diffuse nebulae, then you are already familiar with the techniques for handling dark nebulae. To sketch dark nebulae that are exposed by diffuse nebulae resting behind them, the act of sketching the diffuse nebula will automatically describe the shape of the dark nebula. For a sketch like this, refer to the techniques discussed in the *Sketching a Diffuse Nebula* tutorial.

The process can get a bit more difficult when observing dark nebulae that are revealed by dense expanses of starlight shining from the Milky Way. Telescopic, binocular, and naked-eye views are all capable of displaying the majesty of these objects. There are a couple of ways to approach such a challenge, and both will be discussed in the following tutorials. The first will cover a naked-eye contour sketch and the second will cover a shaded telescopic sketch.

6.3 Producing a Contour Sketch of a Dark Nebula

One way to approach a dark nebula sketch is to bring along a preprinted star chart of the area occupied by the nebula. If you have access to planetarium software, you can print a chart of the region. Or you can make a copy of a page from a star atlas that covers the area in which you are interested. Using this method will allow you to concentrate completely on the shape and extent of the dark nebula itself without being concerned with the position of the multitude of stars that fill the view.

In this example, we will further simplify things by defining the nebula with a contour outline. The subjects of this sketch include the Great Rift that cleaves through Sagittarius and Scorpius and the Pipe Nebula in Ophiuchus. Like most dark nebulae, these require a dark observing site to view.

Step 1: Preparing a preprinted star chart If you are preparing your preprinted star background using planetarium software, there are a couple

Astronomical Sketching: A Step-by-Step Introduction

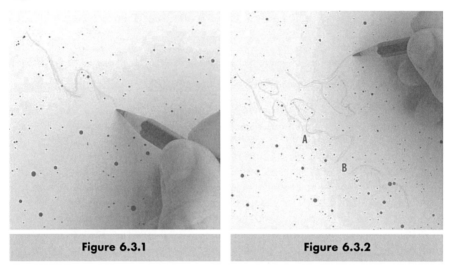

Figure 6.3.1   Figure 6.3.2

of things to consider. First, if you are not familiar with the extent of the dark nebula, you may want to print a few sheets of the region at different scales and bring them all with you so you can use the one that is most appropriate for the observation. Another thing to consider is how large your software plots brighter stars. If a test print reveals bright stars that are distractingly large, see if your software allows you to change the scaling of star weights. If not, you may want to lay a second sheet of paper over the print and redraw the stars to a more reasonable size. If you find it difficult to see through well enough to trace, try holding the sheets together with a small piece of tape and holding them against a sunlit window, a television screen, or a computer monitor. You may be able to apply the same technique to a star atlas that you are photocopying or tracing by hand.

Step 2: Examining the dark nebula Although this tutorial presents a sketch based on a naked-eye observation, these techniques would work just as well with binoculars or a telescope. The first thing you will need to do is orient your preprinted star chart to the sky itself. This may be difficult if you are not intimately familiar with the area, so take your time and make sure you that are equating the correct stars on your plot with what you see in the sky. For a naked-eye observation like this, averted vision is as important as it is with a telescope or binoculars. Take your time scanning the area, and as you begin to detect subtle, dark structures, notice which stars border or bracket them. Find a portion of the nebula that resides near an easily identifiable star pattern or a portion of the nebula that is most obvious and start your sketch there.

Step 3: Roughing in the boundary Use a 2H pencil to lightly feather in the boundaries between the dark nebula and its brighter surroundings. (**Figure 6.3.1**) There may be places where the backdrop to the dark nebula becomes too faint to tell where the nebula's boundary resides. In these places, just let your rough outline taper off as noted by A and B in **Figure 6.3.2**. If you make any significant errors with the outline, the light touch you have used so far should allow these mistakes to be erased fairly easily. Continue to lightly rough in the

Sketching Nebulae

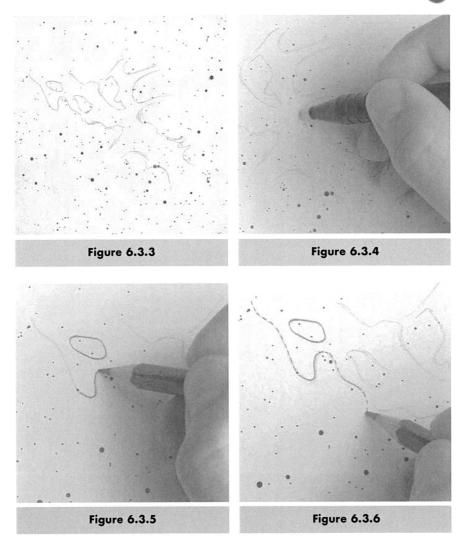

Figure 6.3.3

Figure 6.3.4

Figure 6.3.5

Figure 6.3.6

outline until you are satisfied with the amount of detail you have captured. (**Figure 6.3.3**)

Step 4: Cleaning up the rough outline Take a moment now to erase any exceedingly rough edges in your outline. I prefer to use an eraser pencil for this step, but any eraser will do. (**Figure 6.3.4**) Feel free to use your eraser shield if you have tight areas to tend.

Step 5: Solidifying the boundaries Although you could leave your sketch as is, refining and darkening the boundary lines can give your sketch a more finished look and allow you to see these borders more clearly. Switch to an HB or darker pencil and carefully trace over your rough border. (**Figure 6.3.5**) In areas where the nebula boundary is diffuse, you may want to mark it with a dashed line to indicate this increased softness. (**Figure 6.3.6**) **Figure 6.3.7** shows the completed outline.

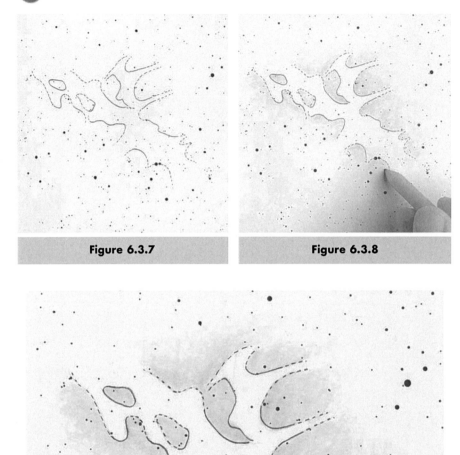

Figure 6.3.7

Figure 6.3.8

Figure 6.3.9

Step 6: Finishing the sketch You can leave the sketch the way it is now. However, if you would like to make it clear which side of the border represents the dark nebula, you can take it one step further. You can label it, or you can use a blending stump loaded with graphite as in the previous nebula tutorials. Using

Sketching Nebulae

even, circular motions, brush graphite on the side of the boundary that represents the luminous backdrop to the nebula. It may be tempting to shade the dark nebula itself, but since this sketch is made in negative form, it is more accurate to shade the luminous regions. (**Figure 6.3.8**) After finishing any notes about the observation, your sketch and observation are complete. (**Figure 6.3.9**)

6.4 Producing a Shaded Sketch of a Dark Nebula

In this example, we will approach a dark nebula sketch by shading it. This will help to convey not only its boundaries but also any subtle structures within the nebula itself. Additionally, it will allow you to depict any variations in brightness within the star field or luminous nebula that rests in the background. You may choose to use a preprinted star field, as in the previous tutorial; however, in this example, the stars will be plotted manually during the observation. For this tutorial, we will make a telescopic sketch of Barnard 34. This cloud of opaque interstellar matter lies a little less than 2 degrees west of the open cluster M37. The observation for this sketch was made from a very dark location with a 15-cm f/8 Newtonian and a 32-mm Plössl eyepiece. This provided a magnification of 37.5× and a true field of view of 88 arc minutes.

Because this dark nebula is silhouetted by the faint, unresolved, starry backdrop of the Milky Way, the observation will not benefit from the use of a nebula filter to increase contrast. My initial impression of the view was of a faint, almost imperceptible "hole" in the softly glowing bed of the Milky Way. More time spent on the observation revealed some extensions of the dark nebula that flowed to the northeast.

Step 1: Framing and preparing the view To begin the sketch, center your view to best show the object or place a convenient star in the center of the view. Mark the cardinal directions around the sketch circle. Then work on plotting the stars that fill the view and provide context for the nebula. (**Figure 6.4.1**) See the *Sketching Star Clusters* chapter for detailed tutorials on this process. With Barnard 34, the view was filled with numerous faint stars that overlapped the unresolved Milky Way beyond. I felt they were very important not only to the aesthetic appeal of the view but also to define the shape of the dark nebula.

Step 2: Analyzing the nebula With these brighter stars plotted, the position of Barnard 34 is already beginning to take shape, but the best definition comes from the foaming

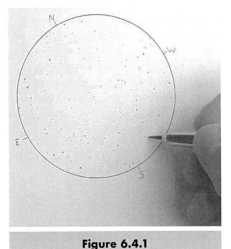

Figure 6.4.1

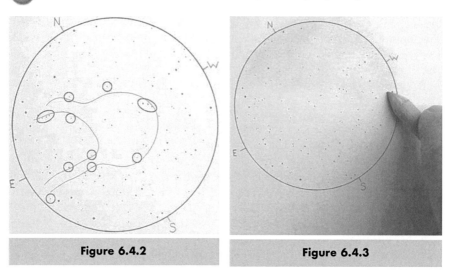

Figure 6.4.2   Figure 6.4.3

Milky Way behind it. Rather than sketching the empty expanse of the dark nebula itself, we will shade in the unresolved starlight and let the nebula take shape by what is not shaded. Using averted vision and slight motions of the telescope can be very helpful in determining the boundaries of the nebula. Use the stars you have plotted as a framework to determine where these boundaries reside. In **Figure 6.4.2**, some of the stars that were key in determining the boundaries of the nebula have been circled. The apparent position of the nebula is marked by a thin line in this image. After getting a good feel for these delineations, find a portion of the view where you would like to start shading. This may be where the relationship between framework stars and the nebula boundary is fairly simple, or you may choose to start in the 12 o'clock position and proceed from there.

Step 3: Applying shading to the background As described in *Using a Blending Stump* on page 153, lightly load a blending stump with graphite. Beginning at the starting point you determined in Step 2, use light circular strokes of your blending stump to feather in the glowing backdrop that surrounds the dark nebula. You can begin at the edge of the sketch circle and work your way inward, or start at the nebula boundary and work outward. (**Figures 6.4.3** and **6.4.4**)

As you shade near the boundary of the nebula, pay attention to how soft or hard that boundary appears in the eyepiece. If it is a soft, gradual boundary, use a very light touch to keep that boundary diffuse.

Step 4: Adding background patchiness The boundary of the nebula is not the only detail to look for. Within the unresolved, starry backdrop, you may discern mottling and larger patches of luminosity. Adjust the pressure on your blending stump to depict these variations. Continue using your frame-

Sketching Nebulae

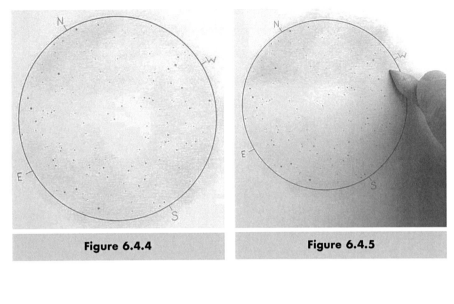

Figure 6.4.4

Figure 6.4.5

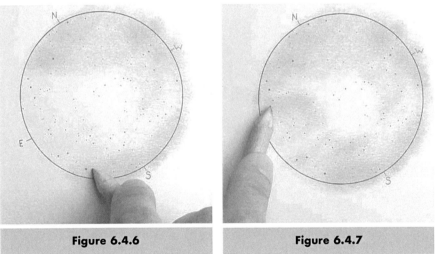

Figure 6.4.6

Figure 6.4.7

work of stars to determine where these areas should be sketched. Do not be afraid to allow the shading to overlap the edge of your sketch circle. (**Figures 6.4.5** through **6.4.7**) As you add heavier shading, you may notice some of your fainter stars becoming blurred to the point of invisibility. If that starts to happen, take a break from shading and lightly mark these stars back in before continuing.

Step 5: Using your eraser After shading this expansive background of starlight or luminous nebulosity, compare your eyepiece view to the sketch. Look for any areas of the dark nebula that have been overreached by the shading

Astronomical Sketching: A Step-by-Step Introduction

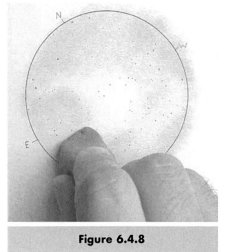

Figure 6.4.8

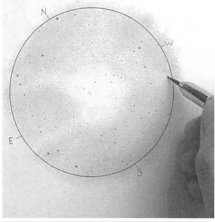

Figure 6.4.9

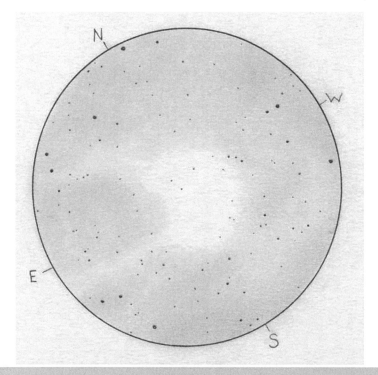

Figure 6.4.10

Sketching Nebulae

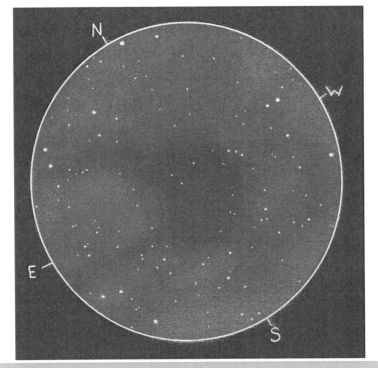

Figure 6.4.11

process. If so, get your kneaded eraser and lightly press or brush it against the sketch to lift graphite away from these areas. (**Figure 6.4.8**)

Step 6: Finishing the sketch Take another look through the eyepiece and compare the relative magnitudes of stars there to what you have sketched. Darken any stars necessary to match the view through the telescope. Take this opportunity to also replot any stars that have been distressed during the blending process. (**Figure 6.4.9**)

If you would like, you can tidy up your sketch by using an eraser to remove any blending that has overlapped the sketch circle. Take a moment to record any notes about the observation, and your sketch is complete. (**Figure 6.4.10**) A positive version of this sketch can be seen in **Figure 6.4.11**.

Tips and Techniques

6.5 Using a Blending Stump

There are several ways to shade a graphite sketch. A pencil can also be used to gently apply shading directly to the paper. Shading applied this way can be

Astronomical Sketching: A Step-by-Step Introduction

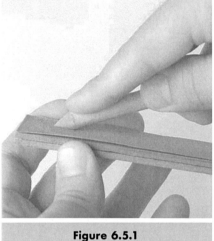

Figure 6.5.1

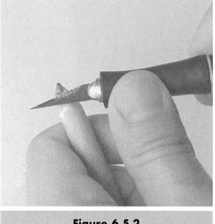

Figure 6.5.2

further blended to achieve a smoother effect by using your fingers, tissue paper, or other materials, including blending stumps and tortillons. Graphite can also be stippled on the sketch to suggest varying shades of gray. In addition to these methods, it is also possible to use a blending stump to apply shading directly to a sketch as though the stump itself were a brush.

This last technique is used frequently in the graphite tutorials I present. I find that applying shading directly with a blending stump provides a great deal of control over the very subtle variations in luminosity displayed by nebulous deep sky objects. The technique discussed here works with charcoal and pastel as well, although the performance of these media will vary compared to graphite. No matter what media you use, experimenting with these methods before you go observing will help you know what to expect when you are at the telescope.

Just like a pencil, a blending stump can be sharpened or sanded to provide a fine point for precise shading. Sanding and sharpening also clean away old used material from the stump and give you a fresh clean surface with which to blend. A sandpaper block can be used to freshen up the stump or hone its point (**Figure 6.5.1**), and an exacto knife can be used to give it a major overhaul (**Figure 6.5.2**). Exercise caution when using a knife to sharpen your blending stump, particularly in the dark at your telescope. In fact, it is probably a good idea to take care of any major sharpening in a brightly lit, comfortable environment before you go observing. If you are uncomfortable using a knife to sharpen the stump, you can still achieve a major resharpening using your sandpaper block. It will simply take longer.

Once your blending stump is prepared, take a soft pencil, such as a B or 2B, and scribble a palette of graphite somewhere outside of your sketch area. (**Figure 6.5.3**) If you would like your sketch sheet to remain clean, you can create this palette on a scrap piece of paper and keep it clipped to the top of your clipboard.

Sketching Nebulae

Figure 6.5.3

Figure 6.5.4

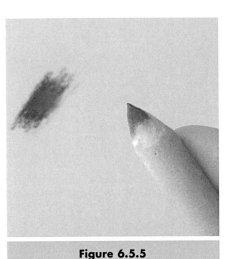

Figure 6.5.5

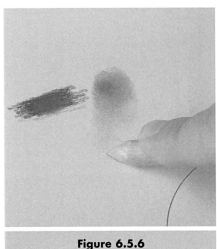

Figure 6.5.6

Next, swirl your blending stump in the graphite palette. (**Figure 6.5.4**) This will cause graphite to adhere to the tip of the stump (**Figure 6.5.5**), but it will probably be too dark if you intend to do subtle shading. To temper this load of graphite, start swirling the stump in a blank area near the palette. As you swirl it, the amount of graphite the stump leaves behind will gradually get lighter. (**Figure 6.5.6**) Continue doing this until the graphite being left by the stump reaches a level that you feel is appropriate for your sketch. Then, holding the stump in the same position, go to your sketch and start to lightly brush the stump where you want to apply the shading. When you first apply it, use a very light touch and

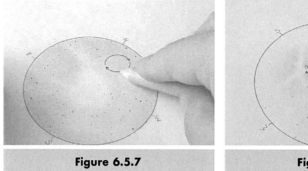

Figure 6.5.7

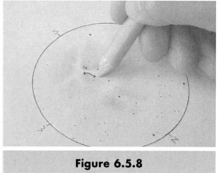

Figure 6.5.8

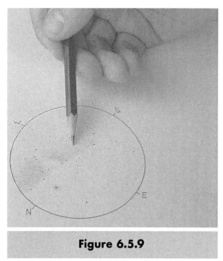

Figure 6.5.9

Figure 6.5.10

gradually increase pressure until the sheet of graphite it leaves on the paper reaches the level of darkness for which you are looking.

When shading with the blending stump, use overlapping circular or elliptical strokes to evenly apply graphite to large areas of your sketch. (**Figure 6.5.7**) When you apply shading to more concentrated areas, adjust the shape of your blending strokes to match the form of the area being sketched. (**Figure 6.5.8**) As you shade, the stump will gradually run low on graphite, at which point you simply need to reload it from your palette and then temper it in a blank area before returning to your sketch.

As mentioned in the introduction to this technique, graphite can be applied directly to the sketch (**Figure 6.5.9**) and blended *afterward* with a blending stump (**Figure 6.5.10**). This method can work very nicely as well. In fact, it can allow you to achieve darker shadows right away that would otherwise require several steps to achieve using the direct blending stump shading technique. There are a

Sketching Nebulae

couple of things to watch out for though. First, with graphite, depending on the type of paper and pencil lead you are using, it is possible that any scratches or swirls the pencil leaves on the paper will not completely blend away. Second, regardless of the medium you use, as you blend the soft boundaries of a nebulous object, the act of blending preapplied graphite may cause those boundaries to expand beyond your original intent. In both cases, practicing the technique ahead of time will help you become familiar with how it performs for you.

6.6 Using a Kneaded Eraser

A kneaded eraser is an extremely versatile tool to keep in your sketching kit. Sold in small squares, they are meant to be kneaded and shaped so you can lift graphite, charcoal, or pastel from your sketch. This eraser is not practical for removing dark, solidly drawn lines. That sort of correction benefits from the use of a pink or other hard eraser. The kneaded eraser is very adept at lifting away subtle amounts of shading, allowing you to carefully control how much you remove. Additionally, since you can shape it like putty, you can broaden or narrow its effects to cover a wide range of needs. As the eraser picks up material from your sketch, it does not leave debris behind. Instead, it absorbs the sketch medium. To freshen it up, you simply knead it and it is ready to be used again. (**Figure 6.6.1**)

To remove small amounts of shading from a broad area, knead the eraser until it is soft and somewhat tacky. Then shape a portion of the eraser to be a bit smaller than the area you want to erase. (**Figure 6.6.2**) Curl this end just a bit so that it is somewhat convex on the side you plan to press against the paper.

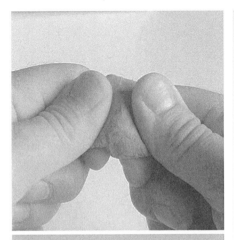

Figure 6.6.1 Freshening up a kneaded eraser

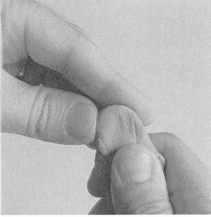

Figure 6.6.2 Shaping the kneaded eraser to pick up a wide area of shading

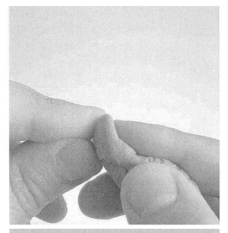

Figure 6.6.3 Curling the base of the eraser to soften the edges

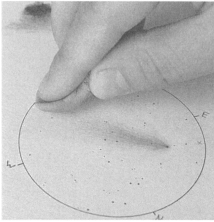

Figure 6.6.4 Lifting a broad, subtle area of graphite away from a sketch

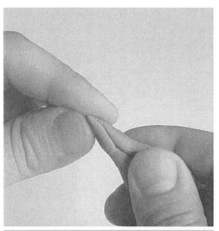

Figure 6.6.5 Shaping the kneaded eraser to pick up a smaller area of shading

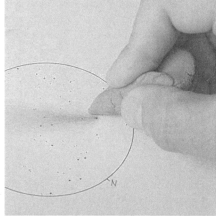

Figure 6.6.6 Removing distinct areas of graphite from the sketch

(**Figure 6.6.3**) Then, using a light dabbing motion, press it against the portion of your sketch that you want to lighten. (**Figure 6.6.4**) The rounded surface and light pressure will help remove the shading and keep the edges of the erasure soft and subtle.

If you want to remove shading from smaller, more distinct, or harder-edged areas, you can shape the eraser into a finer or more chiseled point. (**Figure 6.6.5**) This tip will be a bit sturdier and will allow you to apply pressure to these tighter areas and remove distinct areas of shading using a dabbing or slight brushing motion. (**Figure 6.6.6**)

Sketching Nebulae

Although the kneaded eraser is not good at removing hard objects like plotted stars, it can be used to lighten stars that have been plotted too heavily.

6.7 Sketching Negative Versus Positive

When sketching a deep sky object using graphite or charcoal on white paper, you are producing a negative sketch. This requires that you invert your thinking to consider bright objects as dark and vice versa, both when you are sketching and viewing it later. This is the most common way to sketch the deep sky. However, if you have ever wondered how to produce positive, white-on-black sketches, here are a couple of methods to consider.

Scanning and inverting If you have access to a computer and scanner, you can scan your original black-on-white negative sketch and then invert it with image-editing software to produce a positive version. This is the method I have used with all of the positive versions of my final tutorial sketches. Depending on the type of tools your image-editing software has, you can further adjust the brightness and contrast of the inverted image and even add any color that you noted. Techniques for editing images on your computer are beyond the scope of this book, but some of the Internet links in Appendix C will take you to websites that provide more information on the subject.

Sketching with pastel pencils Another option is to sketch directly on black paper, such as Strathmore Artagain 60# black paper using a white pastel pencil. White Conté, Cretacolor, CarbOthello, and other brands of pastel pencils are available from many art stores and online suppliers. A pastel pencil is soft compared to graphite, so you will need to sharpen it more frequently. Applying pastel shading uses many of the same principal techniques as are used with graphite. The main difference that you will note is with the performance of the materials. You can rub a blending stump on the tip of the pencil to pick up the pastel material and then apply it to your sketch. You can also use your pencil to apply the pastel directly to the sketch and blend it afterward with the blending stump. To get a feel for how this medium behaves compared to graphite, the best thing you can do is to practice ahead of time. See Erika Rix's tutorial, *Ha Filter Sketching: Prominences* on page 58 and Rich Handy's *White Chalk on Black Paper Sketching* Tutorial on page 16 for some great technical pointers on sketching with chalks or hard pastels.

Pastel renderings tend to be more delicate and prone to smearing than graphite sketches. Thus, you will need to take more care when sketching in order to avoid brushing against anything you have already drawn. Because of this and potential visibility problems in the dark, some observers may be reluctant to produce a white-on-black sketch at the telescope. If you prefer, you can sketch black-on-white while at the telescope and then redraw your sketch as a positive later while you are in a more well-lit, controlled environment.

When reproducing the star field, you can visually compare and redraw it or you can use the following tracing method. First, lay a sheet of tracing paper over your original sketch and plot each of the stars on the tracing sheet using a

Astronomical Sketching: A Step-by-Step Introduction

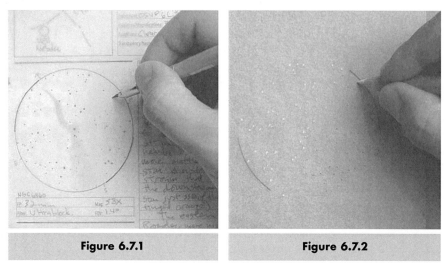

Figure 6.7.1

Figure 6.7.2

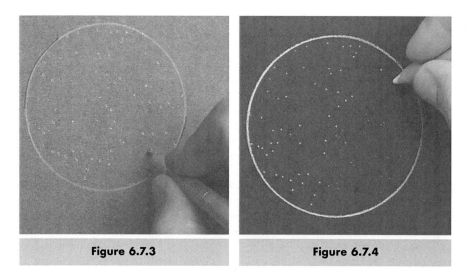

Figure 6.7.3

Figure 6.7.4

graphite pencil. (**Figure 6.7.1**) Sketch a portion of the sketching circle as well for alignment later. Then turn the sheet over and use your pastel pencil to swirl a bold patch of white over each of the stars you marked. (**Figure 6.7.2**) Do not lay the tracing paper over your original sketch while you do this, since you may transfer some of the graphite from the tracing paper back onto your sketch. When you have finished, turn the tracing paper back over and line it up against a circle that you have already drawn on the black sheet of paper. Now, using a sharp graphite pencil, carefully press it against all of your traced stars. Give the pencil

Sketching Nebulae

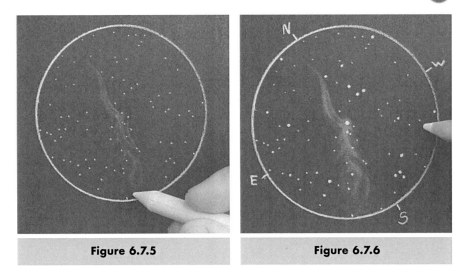

Figure 6.7.5 Figure 6.7.6

about a quarter turn as you press down to be sure that it transfers the pastel onto the black sheet of paper. **(Figure 6.7.3)** Once you are finished, use a sharp pastel pencil to brighten up these ghosted stars and proceed with the rest of your sketch. **(Figure 6.7.4** through **6.7.6)**

CHAPTER SEVEN

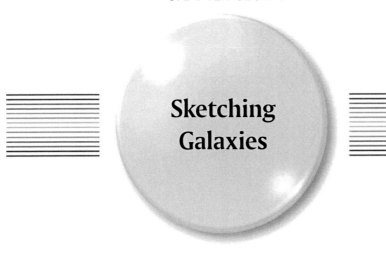

Sketching Galaxies

Galaxies, like other deep-space objects, present some interesting challenges in sketching. They are faint, often very small, and any structure they may exhibit is subtle at best. You really have to be under pretty dark skies to sketch them, which is like reading in the dark or mowing the lawn blindfolded, neither of which makes a lot of sense.

However, it is possible to sketch galaxies, despite how absolutely nuts it might seem. So we will take a look at some of the different methods and techniques for sketching them. After a bit of practice, it may not seem like such a crazy venture after all.

The first thing I can recommend about sketching galaxies is how you get to Carnegie Hall: "Practice, practice, practice." Since you are going to be drawing this stuff in the dark, the process can be much easier if you learn to comfortably draw galaxies in the light.

You will see several references throughout this chapter about drawing in the dark. A lot of what will be discussed is in that context. When you start trying these techniques out in the field, think about how you will do them under minimal light. I have also included some tips and tricks at the end of the chapter, many of which I have found especially useful out under the dark skies.

Astronomical Sketching: A Step-by-Step Introduction

7.0 Getting Started

Before we practice, a couple of comments are in order about our tools of the trade. Generally, what you need for sketching galaxies is pretty simple: a pencil or set of pencils (graphites), charcoal, plastic eraser (a gum eraser can be handy as well), eraser shield, a few blending stumps (large and small), and perhaps a chamois. Additionally, you will probably want some paper, unless, of course, you prefer drawing galaxies on the back of your hand. (**Figure 7.0.1**)

Now, for this chapter, we will be drawing black on white (an inverted image or in the negative). I generally prefer doing it this way because it is easier to see while drawing in the dark. You can later scan this "negative" into a computer and invert it so that it will look like what we see in the eyepieces (white on black).

A great place to start practicing on galaxies is to play with your drawing tools, especially the charcoal and blending stumps. They will be your primary tools for creating your island universes. Just make some different shapes and curves and sort of play around so you can get the feel of what you can make. (**Figure 7.0.2**)

For some more serious practice, you will also want to start with galaxy models. Fortunately, for galaxies, we happen to have an excellent drawing model to work with: the Hubble "tuning fork." This was Hubble's visual morphological categorizing of galaxies, which works out very well for us since this is a very visual thing we are doing. The Hubble tuning fork has the elliptical galaxies on the left and the spiral and barred spiral galaxy shapes on the right. The lenticular galaxy shape is in the middle. (**Figure 7.0.3**)

This represents some, but certainly not all, of the major galaxy shapes that you might commonly encounter. There are many other divergent shapes out there, but these are several great shapes with which to work and practice. Drawing these

Figure 7.0.1.

Sketching Galaxies

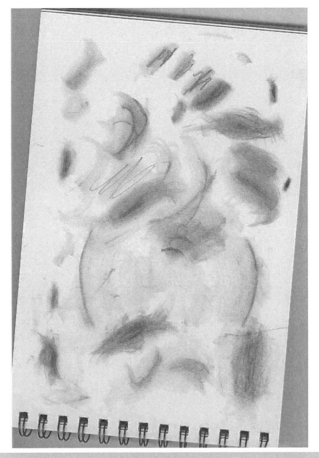

Figure 7.0.2.

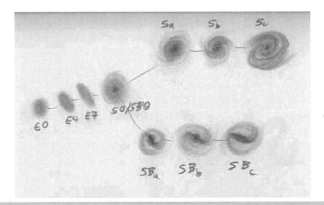

Figure 7.0.3.

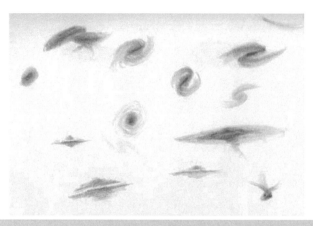

Figure 7.0.4

shapes and structures comfortably will make grasping the other more peculiar shapes easier. (**Figure 7.0.4**)

Practice is where you start, but you will never finish practicing. I constantly doodle fictitious galaxies on sketching paper, trying to make those spiral arms look just right, that dark lane more natural, and getting that elliptical to fade out on the edges in just the right way.

7.1 Your First Galaxy

Now, many you reading this may want to immediately start drawing something like the Andromeda galaxy (M31) or something equally ambitious. Although that is a great subject for many people starting into astronomical imaging, it really does not work well for most beginning astronomy sketchers because it is very complex and far reaching.

Instead, we are going to start with one of my favorite barred spiral galaxies, M66. This bright galaxy is in Leo and is part of the well-known "Leo Triplet." It makes a great subject to sketch because it has an interesting barred structure that is somewhat on its side. However, for beginners, it is not overly large and complex, so it will be fairly easy to take in visually and learn some basic techniques. You can see it very easily in smaller apertures.

First, let us sketch a star pattern. It is a chicken-or-egg thing whether you want to put the stars in first or the galaxy (or galaxies). I generally prefer drawing the stars first because they help me frame and scale the galaxy in my drawings. However, usually I just limit it to the brighter stars, putting the really faint stars in last. (**Figure 7.1.1**)

You will notice that my pencil is fairly straight up and down. I will move it down to the paper, pressing it only enough to get the star how dark (remember,

Sketching Galaxies

Figure 7.1.1

Figure 7.1.2

dark = bright here) I need it to be. Sometimes, to get a bit darker stellar dot, I will slightly spin the pencil in my fingers, maybe a quarter turn. It is really important that you keep a somewhat sharp point on your pencil (keep a sharpener handy) with just a little bit of smooth bluntness on the end.

Now that we have the brighter stars in place, it is time to put the beginning of the galaxy into place. This is where the charcoal is used and there are a variety of ways to apply it.

Usually, I take my charcoal and draw a dark well of black in the corner of the paper. It is from this well that I will pick up the charcoal with my blending stump and apply it to the paper for the picture. In **Figure 7.1.2**, you can see the dark well near the top of the picture.

Another option is to apply the charcoal directly. I do not do this for most subjects, but there are times, such as when I have a very distinct core for a round elliptical, that I will want to use that technique. A third option is to apply or rub a bit of charcoal to the end of the blending stump and then lay it on the paper. This is an especially useful technique for drawing a faint filament or arm from a galaxy.

Probably one of the most important things to remember about applying charcoal with a blending stump is that it starts off dark and gets lighter (something we will call "charcoal fade"). How quickly it gets lighter will depend on how hard you press the stump. The charcoal applied will get lighter fairly quickly in any case. This means that you should plan your charcoal lines accordingly.

For instance, when drawing a spiral galaxy arm, you should start at the bright core (which would be drawn dark) to the fainter outer arms. As much of this that you can do in one stroke will give the arm more continuity. If you are encountering a mottled galaxy, then short, little strokes outward from the brightest parts of the mottling are probably more in order.

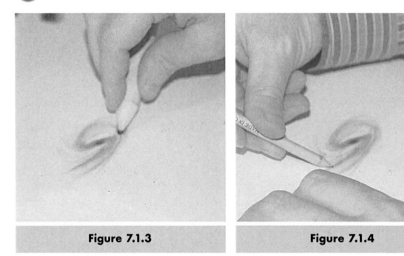

Figure 7.1.3   Figure 7.1.4

For M66, I am using the stump to draw out the spiral arm. You can see that I am using the tip of the stump in the direction of the end of the arm so I can finish the arm with a nice soft point. (**Figure 7.1.3**)

The continuity of charcoal fade can also be controlled by the size of the blending stump. For larger galaxies, or long, broad spiral arms, a large blending stump is appropriate. However, as you can see in **Figure 7.1.4**, I am using a small blending stump to finish in some small details on the major arm of M66. M66 has some really neat dark lane stuff going on in its forward arm. It shows up nicely at high magnification in medium to large apertures, which was how this sketch is being drawn.

The last step I usually take in any sketch is to capture the faintest items. These are usually the faintest stars, although I sometimes find myself getting a bit of faint nebulosity or a tiny galaxy that I did not get before. I typically spend a great deal of time just looking into the eyepiece and asking myself, "What did I miss?" This is can end up being your most important step, because you very well may find that you missed something obvious.

In **Figure 7.1.5**, I am putting in the final faint stars with a harder graphite pencil. I am stippling, as always. This requires the finest and lightest touch. These stars are at the very edge of what I can observe through my telescope. I use the brighter stars and galaxy structures as my road map for placing them.

Something you should always remember is this: The eraser is your best friend. The eraser's best friend is the eraser shield. If you draw much in the dark, you will always have accidental stray marks and smudges. Or you may accidentally misplace a star while drawing (I will show you how I handle that in the tips and tricks at the end of the chapter).

The eraser shield helps you eliminate such errors and problems while protecting those things that you do not want to erase. You may notice in **Figure 7.1.6** that I am using a plastic eraser, as opposed to, say, Art Gum. I find that with charcoal smudges and darker stars, the plastic stick eraser with the shield is a quick and effective way to make corrections, especially in the dark. My primary use of an Art Gum eraser is to reduce the darkness of a charcoal stroke or to take

Sketching Galaxies

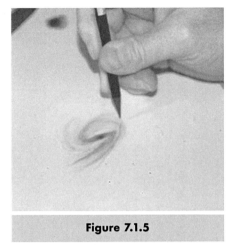

Figure 7.1.5

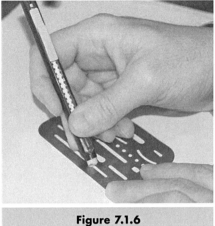

Figure 7.1.6

Figure 7.1.7

the magnitude of a star down a peg or two. Art Gum can also be very useful with erasing the fainter elements, but I find that its use in the dark is not easy for making precise erasures.

Here is your finished M66. (**Figure 7.1.7**)

Figure 7.1.8

And here is the inverted image. (**Figure 7.1.8**)

7.2 The Next Step: An Irregular Galaxy and Dark Lanes

The next thing we are going to sketch is an irregular galaxy: M82, the Cigar Galaxy. This one is a bit more challenging because it does not visually fit into the regular patterns of our Hubble tuning fork, although it does have some elements found in the different classifications. M82 also has an unusual dark lane pattern in front of it.

Dark lanes in galaxies (and other nebulae) represent a technique worth practicing and will take a little patience to master. What you are trying to do in the inverted images (such as we are doing here) is trying to make an area lighter in a galaxy (thus, darker in its positive counterpart). The challenge is that a dark lane is usually a very narrow and well-defined region. The two mistakes occurring most often are that the dark lane either ends up too light or too big.

So, first, practice, just as I mentioned in the first part of this chapter. Second, use an easy touch here. You do not want to completely erase an area, just lighten it enough to give a visible contrast. Third, keep and use a sharpened edge of an eraser to do this. This is a technique worth mastering since you will encounter it many times in astronomical sketching.

Sketching Galaxies

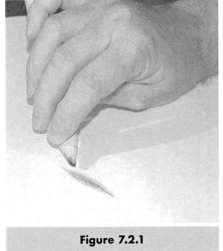

Figure 7.2.1

Figure 7.2.2

M81 and M82 are found by each other in Ursa Major. M81 is a very graceful spiral, whereas M82 looks like its cigar namesake after someone made it explode. The dark lanes going down the middle of M82 lends to a helpful rhyme that is useful in remembering which one is which: "M Eighty-two, split in two."

We are also going to take a different approach with M82. Instead of putting the bright stars in first, we are going to draw the galaxy first. This is an easy choice here because M82 really dominates its field, so everything else in the field of view can be defined by it.

Both the large and small blending stumps will be used aggressively here. The large blending stump will be used for the major dimensions and structure, whereas the smaller one will be useful for better defining the edges to the dark lanes and the galaxy's mottling.

Now, looking through the telescope, the first thing I notice about M82 is that although its center is bright compared to its edges, it really does not have a strong, distinct core. Part of this is because of the large dark lane through the middle. However, generally speaking, this galaxy is fairly bright right up to the edge, giving it a well-defined cigar shape.

It is that bright cigar shape that we will start with as our base. First, we will use our charcoal and blending stump and apply an even band outward from the center, tapering off the ends. (**Figure 7.2.1**)

We will now develop the body a little better, giving a gradual fade to the edges. You want to be careful, however, and not overdo this, otherwise you will find the charcoal spreading out much farther than the actual size of the galaxy that you see in the eyepiece. This is where the smaller blending stump becomes very useful. You can also use a more rounded tip (see tips and tricks at the end of this chapter) of a smaller stump to mottle the texture of the surface brightness of this galaxy. (**Figure 7.2.2**)

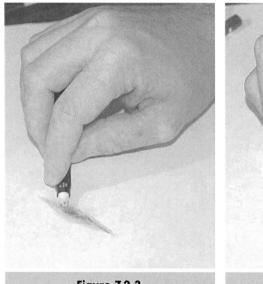

Figure 7.2.3

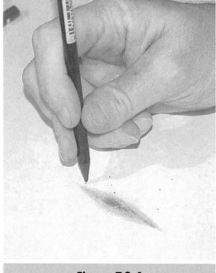

Figure 7.2.4

The dark lane bisecting M82 is most easily done at this point with a plastic eraser, using its edge, gently applied just off center through the galaxy. Remember, in this instance you are not completely erasing the charcoal you already applied, just lightening it enough to give a significant contrast. (**Figure 7.2.3**)

The galaxy is finished; all you need are the field stars. (**Figure 7.2.4**)

Now you are done! (**Figure 7.2.5**)

7.3 Tips and tricks

This chapter represents a first step into sketching galaxies. I used a couple of different examples of galaxies to bring out some of the more frequently used techniques that you might need while drawing in the dark. However, here are a few other nice little tips and tricks that might be useful as you further your sketching.

Woodless Graphites for Drawing in the Dark

I generally like to use woodless graphites, especially on stars, because they are heavier than normal wooden graphites (pencils) and that makes them easier to feel and control in the dark.

Sketching Galaxies

Figure 7.2.5

Use an Eye Patch

I typically use one eye for sketching and the other for observing. The observing eye I keep under an eye patch to maintain my dark adaptation.

Illuminate Your Sketching Surface Wisely

Use a diffused, wide-angle, evenly distributed low red light for illuminating your sketching surface. Use the lowest setting you can and still see the sketching surface. A wider, even glow is much preferred over a narrower spot because it makes it easier to see the different shadings of galaxies and magnitudes of stars as you draw them.

Using your Eraser Shield for Certain Dark Lanes

This is great for galaxies like the Sombrero Galaxy (M104), which has a well-defined straight or near straight dark lane going down its length. The typical eraser shield has some straight lines and slightly curved lines cut into its template. Applying these lines over the body of a galaxy with an eraser can yield a great dark lane. (**Figures** 7.3.1 and 7.3.2)

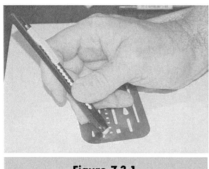

Figure 7.3.1

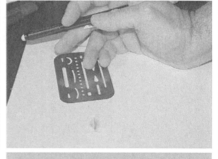

Figure 7.3.2

The curved-line templates can be especially useful with dark lanes you might find in "face on" spiral arms. As always, this takes practice to master, so play around with the technique before you take it to the field.

Blending Stump Tips

Keep one end of the blending stump sharp and the other a little dull. This will make it easier to alternately create sharp, soft, fine, and broad features with the stump.

Extra Stumps

Keep an extra blending stump with a clean tip available for the end of a sketch. Sometimes at the end of a sketch, you will have charcoaled areas that you want to lighten up just a bit, but an eraser may take away too much. This is where a clean blending stump can come in handy. The clean (or freshly sanded) surface of the stump will pick up excess charcoal and lighten an area or give a subtle light line.

Blending with Graphite

Blend a galaxy with soft graphite instead of charcoal. I do this for very light and small galaxies. I prefer charcoal in most instances because of its greater dynamic range. However, sometimes, with the fainter galaxies, you do not need or even want that type of range. Here is where a soft graphite pencil can be easier to control, especially in the dark. You may want to soften the paper surface a bit though, to dull the sheen of the graphite. This can be done by rubbing a rounded blending stump tip against the paper just a little bit harder than you normally would. The rubbing will break up the surface fibers of the paper just enough to flatten the surface. Be careful not to do this too hard or your drawing surface will give unreliable results.

Correcting Misplaced Stars

Sometimes while sketching at night, I will immediately notice that I just misplaced a star position. Instead of trying to erase it in the middle of a sketching process, I will just draw a faint little arrow from the star to where it should be. The end of the arrow will indicate the correct location. This will make it much easier to correct later at any time.

7.4 Some Great Starter Galaxies to Sketch

Here is a short (and not all inclusive) list of some good galaxies to use as your first galaxies. They will give you practice over a variety of techniques while remaining manageable. They will also be great ones to revisit later.

- M81 (Bode's Nebula): This galaxy's challenge is in the subtleties of its fine and broad arm structures.
- M82 (Cigar Galaxy)
- M104 (Sombrero Galaxy): The dark lane through the center is something you will want to try over and over again.
- M66 (barred spiral)
- M84 and 86 (elliptical): These are two excellent bright elliptical galaxies next to each other in Virgo.
- M100 (elliptical): This is a great elliptical in Leo.
- M64 (Black Eye Galaxy): This has a wonderful short dark lane feature.
- M109 (barred spiral): Distinctive bar structure. It can be a bit of a challenge to see with a nearby bright star.
- NGC 1365 (barred spiral): The bar on this bright galaxy in Fornax gives it a very striking shape in the eyepiece, making it a great sketching target.
- NGC 891 (edge on spiral): This popular galaxy in Andromeda has a dark lane along its middle and can be a good challenge to get just right.

There are many others, of course. The great thing about finding galaxies to sketch is that you will never run out of different and interesting targets.

Now go forth and sketch!

APPENDIX A

Observing Forms and Sketch Templates

178   **Astronomical Sketching: A Step-by-Step Introduction**

Lunar Sketching Form

Date:	Time Started:	Time Ended:
Seeing (Antoniadi):	I II III IV V	(circle)
Weather Conditions	Temp: Dew Pt.:	Humidity:
Telescope:		Focal Length:
Eyepiece(s):		Barlow:
Lunation:	Phase:	Colongitude:
Atlas Page:		
Lunar Feature:		
Notes:		

Figure 8.1.1

Observing Forms and Sketch Templates

Figure 8.1.2 Sketch template for Saturn

Figure 8.1.3 Sketch template for Jupiter

Observing Forms and Sketch Templates

Deep Sky Observing Form

Subject:			
Date:	Time:	Observing Location:	
Instrument:		Aperture:	Focal Length:
Eyepiece:		Filter:	Magnification:
Conditions:			True Field of View:
Seeing:		Transparency:	

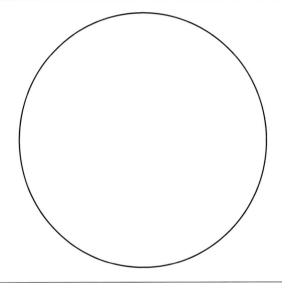

Notes:

Figure 8.1.4

APPENDIX B

Glossary

Antoniadi Scale: Scale used to record atmospheric turbulence, devised by E. Antoniadi. It is mainly used for planetary and lunar observations.
  I. Perfect seeing with no quivering
  II. Slight quivering with moments of calm
  III. Moderate seeing
  IV. Poor seeing with constant tremors
  V. Very poor seeing to the point that it is near impossible to create a sketch
The Pickering can be used as an alternative: 1 (very poor) through 10 (excellent).

Art Gum Eraser: Soft rubber eraser often used for pencil work; has a larger crumbling grain than a vinyl eraser, but can give a broader depth or dynamic range to the level of erasure; very useful for erasing broad areas or where only a soft or partial erasure is desired.

Blending Stump: Round stick of pressed paper with pointed ends, used to blend or spread charcoal, chalk, and graphite across the sketching paper.

Carpentry Pencil Sharpener: These plastic sharpeners can be purchased at hardware or lumber supply stores. They have wide openings that allow the flattened lead carpentry pencils to be easily sharpened to a point. They are great for sharpening the Conté Crayon sticks as well.

Chamois: Soft leather cloth used for blending sketched markings, held by stick tools or fingers. Its soft, fine texture produces an even blend.

Charcoal: Sticks of charcoal, available in various degrees of tone, used to create a broad dynamic range of gray–black. Charcoal generally has a much larger grain than graphite (pencils) and yields a flatter marking, producing fewer glares from sketching lights at night. There are "white" charcoals that are similar to pastels, but behave like charcoal in their application to paper.

Charcoal Holder: Mechanism used to hold round charcoal sticks with pencil like control.

Colored Pencils: Pencils in a variety of colors used alone or in conjunction with chalks for colored renderings.

Compass: Device used to draw circles and arcs.

Conté Crayons/Pencils: Chalks and pencils in assorted colors and tones with rich textured results, invented by Nicolas Jacques Conté in the late 1700s during a

Glossary

shortage of graphite. He combined powder graphite with clay to create Conté pencils and crayons. Both chalk and pencils are softer than regular chalks and pencils, and the pencils require sharpening more often than graphite. Conté pencil renderings tend to be more delicate and prone to smearing than graphite sketches; however, the Conté Crayons seem to hold much better than regular chalk. Blending stumps can be applied directly to both the crayons and pencils to blend and shade.

Cosmetic Brush: Soft brush used for cosmetics that is particularly useful in brushing away eraser debris from sketching paper.

Dry Cleaning Pad: A bag filled with fine inert powder that is useful for cleaning dirt and smudges off sketching paper without abrading the surface.

Easel: Platform made of wood, plastic, or metal that holds the sketch board while drawing.

Eraser Shield: Thin metal sheet with small precision-tooled openings that is useful for making precise erasures while protecting the rest of the sketch.

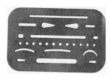

Exacto (X-ACTO/Utility) Knife: Precision cutting knife used to sharpen graphite, chalk, charcoal, and erasers.

Felt-Tipped Pens: Hard acid free felt tipped pens that are available in a wide range of precision tip diameters measured in fractions of a millimeter. These are very useful for stippling and can be used in lieu of technical pens or India ink and quills.

Gradation: A barely perceptible transition of tone or color value.

Graphite: Essentially pencil "lead" available in many different degrees of tone, encased in wood (just like normal writing pencils) or "woodless," where the whole pencil body is graphite with a thin plastic coating over it. Graphite usually has a smaller dynamic range than charcoal, but a much finer grain. This often gives a softer, shinier appearance and the ability to add much finer and fainter detail.

Hubble Tuning Fork: Visual classification of galaxy shapes (morphology) created by Edwin Hubble. It includes the most common galaxy types: elliptical, lenticular, spiral, and barred spiral. This classification does not include the more peculiar and irregular galaxy formations. It is called the tuning fork because of its shape.

India Ink and Quill: Intense black ink used on absorbable media by dipping a brush or a quill into the ink to create broad strokes, washes, and precise details. The quill consists of metal tip (available in various stroke sizes) held by a wooden or plastic handle.

Kneadable (Kneaded) Eraser: Soft pliable rubber compound (many modern types are plastic) that can be shaped and kneaded. Typically, it is a medium grain eraser (although the plastic types are finer). This eraser can be used to lift smudges and marks with minimal abrasions to the sketching surface. Any residue falling off during an erasure can be rolled right back into the eraser.

Negative Sketching: Technique of sketching light features as dark, and visa versa. An example is DSO sketching. If white paper is used, dark graphite or charcoal would represent the stars and DSOs, whereas through the eyepiece the background would be dark and the stars would be light. This technique can be a little tricky at first, especially with nebulae and lunar features, but it does become easier with practice.

Rite in the Rain Paper: All-weather writing/sketching paper that remains very durable under extremely damp conditions. This paper is excellent for high-humidity areas where normal sketching paper becomes too soggy. It has a very fine tooth, and although it may smear a little using charcoal, it is very minimal compared to most other papers. Graphite blends wells, but holds a clean mark instead of smearing. The only downfall is that it is difficult to get a clean erasure; however, vinyl erasers do seem to work the best with this medium. If your sketching technique involves using layers where you rarely require the use of an eraser, this paper is very suitable. (www.riteintherain.com)

Glossary

Sandpaper Block: Layers of fine sandpaper on a flat stick that is used for sharpening and cleaning tools such as blending stumps, tortillons, pencils, charcoals, and pastels.

Sponge: Any type of natural or synthetic sponge used for blending broad sketched strokes. They can be cut down to various sizes and come in a wide range of textures.

Stealth Bomber Pencil Sharpener: Combines ingenuity and stealthlike capabilities while sharpening pencils in the dark.

Stippling: Sketching technique of producing randomly positioned dots or lines of black ink so that spacing density will be perceived as various gray tones to the eye. Stippling also refers to the techniques used for making stars and star patterns in astronomical sketching.

Strathmore Paper: Acid-free recycled paper that comes in a variety of textures and weights and is ideal for pastels and charcoals; made with over 50% recycled materials.

Technical Pen: Type of drawing pen often associated with engineering and architectural drawings. It has two distinct characteristics: (1) The drawing tip and the associated line width it draws has a fixed width, usually measured in decimal fractions of a millimeter and (2) the ink uses a flow delivery (similar to a fountain or felt tip pen) rather than a ball delivery (such as a ballpoint pen).

Tortillon: Stick of tightly rolled paper/cardboard with a pointed end that is useful for blending very fine details. It is available in various sizes.

Vinyl (Plastic) Eraser: Eraser made of soft plastic with a much finer grain than an Art Gum eraser. It often gives a very clean erasure of pencil, chalk, and charcoal marks with very little remaining residue (if any). This type of eraser is especially useful when total erasures of small detailed areas are required and is available in pencil form that can be sharpened with an Exacto Knife.

APPENDIX C

Online Resources

Astronomical Sketching Communities

http://www.cloudynights.com/ubbthreads/postlist.php/Cat/0/Board/Sketching
  Cloudy Nights Telescope Reviews—Sketching Forum: friendly community of astro-sketchers discussing techniques, tools, and advice

http://groups.yahoo.com/group/astrosketch/
  Yahoo Astro-sketch website

Lunar

http://aa.usno.navy.mil/data/docs/RS_OneDay.html
  Complete Sun and Moon Data for One Day

http://www.inconstantmoon.com/
  Inconstant Moon

http://www.moonsketch.com/index.html
  Lunar sketch gallery by Rich Handy

http://cityastronomy.com/
  Whitepeak Lunar Observatory: superb amateur site devoted to lunar geologic studies by Mardi Clark

http://www.lpod.org/
  LPOD Lunar Picture of the Day

http://nssdc.gsfc.nasa.gov/planetary/planets/moonpage.html
  NASA Lunar and Planetary Science: The Moon

http://www.astrosurf.com/avl/UK_index.html
   Virtual Moon Atlas by Christian Legrand and Patrick Chevalley: freeware for all levels of lunar study and observation planning

http://www.cloudynights.com/item.php?item_id=1273
   Virtual Moon Atlas Tutorial for beginners on Cloudy Nights by Erika Rix

http://www.psrd.hawaii.edu/Aug00/newMoon.html
   PSRD: A New Moon for the Twenty-First Century, by G. Jeffrey Taylor: geological study

Comets

http://solarsystem.nasa.gov/planets/profile.cfm?Object=Comets
   NASA: overview and gallery

http://cfa-www.harvard.edu/iau/Ephemerides/Comets/
   Harvard: Orbital elements and ephemerides of comets that are or may have potential to be observable

http://www.nineplanets.org/comets.html
   The Nine Planets—Comets by Bill Arnett: information, history, links, images, press information, including a recipe to make your own comet

http://www.eaas.co.uk/news/ visual_comet_hunting_2.html
   Why Observe Comets by Martin McKenna, EAAS

Solar

http://www.sungazer.net/index.html
   Sungazer.net: discover, explore, learn about the Sun: includes comprehensive resource and equipment information by Greg Piepol

http://www.lpl.arizona.edu/~rhill/alpo/solar.html
   ALPO Solar Section

http://www.bbso.njit.edu/
   Big Bear Solar Observatory

http://spaceweather.com/
   Daily update of space weather news, excellent for planning and verifying solar observations

http://www.aavso.org/observing/programs/solar/
   AAVSO Solar Observing Program

Planets

http://www.lpl.arizona.edu/alpo/
   Association of Lunar and Planetary Observers

http://skytonight.com/observing/objects/javascript/mars/
   Sky and Telescope's Mars Profiler

http://tes.asu.edu/
   Arizona State University: Mars global surveyor

Online Resources

Deep Sky Objects

http://www.seds.org/messier/
SEDS, The Messier Catalog

http://www.noao.edu/image_gallery/galaxies.html
NOAO Image Gallery

http://www.ngcic.org/
NGC/IC: terrific resource for image and object data

http://web.mac.com/bicparker
Deep sky sketches by David Moody

General Astronomical Interest

http://www.astronomicalsketching.com
Astronomical sketching resources by the authors of *Astronomical Sketching: A Step-by-Step Introduction*

http://www.cloudynights.com/
Cloudy Nights Telescope Reviews: comprehensive, friendly community of amateur astronomers

http://beltofvenus.perezmedia.net
Astronomical sketches and observations by Jeremy Perez. Features sketching tips, techniques and tutorials; printable observing form/sketch templates; and links to other online astronomical sketching resources

http://www.astroleague.org/
The Astronomical League

http://www.astrosociety.org/index.html
Astronomical Society of the Pacific

http://solarsystem.nasa.gov/index.cfm
NASA: Solar System Exploration

http://www.slackerastronomy.org/wordpress/index.php
Slacker Astronomy with excellent Podcasts

http://www.astronomytogo.com/
Astronomy to Go, education outreach organizations and home of the Yard Scope

http://cleardarksky.com/csk/
Clear Sky Clock Homepage

http://scienceworld.wolfram.com/astronomy/
Eric Weisstein's World of Astronomy

Index

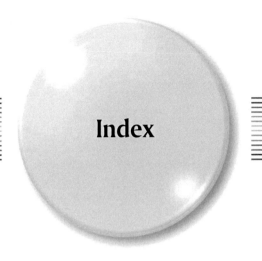

A
73P-C/Schwassmann-
    Wachmann 3   24–32
A Ring   74
albedo
    Jupiter   84
    Mars   86–88
    Saturn   76
aligning, solar scope   50, 54
Antoniadi scale   Glossary
archiving   41–42
averted vision, planetary   70

B
B Ring   72, 74
Barnard 34   149–153
Barnard 85   135
binocular sketching   32–38
blending
    Jupiter   80
    Mars   90
    Saturn   76
blending stump (see tortillon)
    3, 26–30, 34–37, 109–111,
    114–116, 119–122, 136–138,
    142–143, 148–151, 153–157,
    159, 167, 168, 171, 174

blending stump rubbed in
    charcoal, pastels, and
    graphites   6–10, 50–51,
    53–55, 58–60

C
C/2004 Q2 Machholz   32–39
cardinal directions, noting
    39–40
Carnegie Hall   (see practice)
Cassini's Division   77
chalk   (see "Conté" and
    "pastels")
chalk, solar sketching   58–68
charcoal
    galaxy sketching   167, 171,
        174
    lunar sketching   6–11
    solar sketching   50, 53–57,
        58, 62
chromosphere
    Disparition Brusque
        48–49
    filaments   57, 66–67
    full disk   63–68
    hydrogen alpha   49, 57–68
    limb darkening   65

chromosphere *(continued)*
    negative sketching   62
    observing tips   49, 58
    plage   57, 65–67
    sketching tutorials   57–68
    sunspots   67
Cigar Galaxy   170, 173
clipboard   2, 43–45, 71
comets   23–39
    73P-C/Schwassmann-
        Wachmann 3   24–32
    C/2004 Q2 Machholz   32–39
    coma   26, 28–30, 34–35
    field stars   25–26, 32–33
    framing   25, 32–33
    noting motion   30–31
    observing   24
    pseudonucleus   28–30, 34
    tail   26–28, 35–37
compass   40
Conté
    deep sky objects   159–161
    lunar   16–21
    solar sketching   58–61,
        63–68
corona   68
Crepe Ring   76, 77

Index

D
dark adaptation   42–43, 58
dark lane(s)   170, 172, 174
dark nebulae   145–153
　Barnard 34   149–153
　Barnard 85   135
　contour   145–149
　field stars   149
　framing   145–146, 149
　Great Rift, Pipe Nebula
　　145–149
　mottling   150
　observing   145, 146,
　　149–150
　shading   148–153
diffuse nebulae   134–141
　field stars   135, 136
　framing   135
　M20   135–141
　observing   134–135
　shading   136–138
Disparition Brusque   48–49

E
Enke Minima   74
eraser
　art gum   2, 168
　kneaded   28, 30, 110, 111,
　　120, 122, 124, 138, 143,
　　151–153, 157–159
　Pink Pearl   2
　plastic   168, 172
　Sanford Magic Rub   2
　Staedtler Mars plastic   2
　use for planetary sketching
　　71, 94
eraser shield   4, 5, 80, 84–85,
　　128, 168, 174
erasing   28, 30, 37, 101, 111, 114,
　　120, 122, 124, 128–129, 138,
　　143, 147, 151–153, 157–159
Eskimo Nebula   141–145
exacto knife   53–55, 128–129,
　　154
exit pupil   70
eye patch   173

F
faculae   54–57
festoons   78, 80
filaments   57, 66–67
filing   41
filters, effect on nebula
　　sketch   135, 136, 138,
　　141, 149
filters, narrowband solar (see
　　hydrogen alpha)   57

filters, white light solar   50,
　　52–54, 57
fixative (see Krylon Clear
　　Matte Finish No. 7120)
　　6, 41–42
floaters   70
following limb, planetary   69
full disk, chromosphere
　　63–68

G
geometric relationships
　　101–102
globular clusters   113–125
　core   114, 116, 119
　field stars   114, 118
　framing   114, 118
　halo   115, 120
　M13   113, 118–125
　M75   113–117
　mottling   120–122
　observing   113–114, 118
　stippling   122–124
　unresolved stars/
　　granularity   115–116,
　　119–125
granulation, photosphere   53
graphite
　palette   154–155
　pencil   126–127
　woodless   172
Great Red Spot   78, 80
Great Rift, Pipe Nebula
　　145–149

H
Hubble tuning fork   164, 165
hydrogen alpha solar filter
　　49, 57–68

I
inverted image, galaxies
　　164, 170

J
Jupiter   78

K
Krylon Clear Matte Finish No.
　　7120 (see fixative)   6

L
Leo Triplet   166
light source (see sketch
　　lighting)
　for observing   42–45
　gooseneck   44–45

light source *(continued)*
　headlamp   45
　LED headlight   2
　musician's sheet music
　　light   71
　red flashlight   42–45
　red flashlight, distracting
　　light pattern   42
limb brightening, planetary
　　94
limb darkening
　Jupiter   84
　Saturn   77
　Solar   53–54, 65
lunar longitude   1

M
M13   113, 118–125
M20   135–141
M29   99–105
M50   105–113
M66   166, 169
M75   113–117
M81   171
M82   170, 173
M104   174
magnification, planetary
　　70
Mars   86
Milky Way   99, 105,
　　145–149
mistakes, correcting   28, 37,
　　101, 111, 114, 120, 122,
　　124, 128–129, 138, 143,
　　151–153
mottling, galaxy sketching
　　171

N
narrowband solar filters
　　57
nebulae (see diffuse nebulae,
　　planetary nebulae, and
　　dark nebulae)   133–153
nebulae, observing   133
negative sketching
　deep sky   159
　galaxy   164
　solar   62
NGC 2392   141–145
Northern Polar Hood   94

O
observing logs   40–42
observing templates
　deep sky   181
　lunar   178

Index

observing tips, solar
  dark adaptation 58
  scanning the solar disk 49
open clusters 98–113
  cluster stars 101–103, 107–108
  defocusing 106
  field stars 100–101, 106, 108–109
  framing 99, 105
  M29 99–105
  M50 105–113
  observing 98–99, 105, 109
  unresolved stars/granularity 109–111
outline, lunar sketching 3, 6

P
paper (see Rite in the Rain/Strathmore)
  acid content, weight, texture 41
  acid-free 12
  affects of weather on 6
  black 58–68
pastels (see chalk) 159–161
pencils
  charcoal 6–11
  color/pastel 71, 159–161
  colored Saturn 77
  colored Sun 62–63, 66–67
  graphite 2, 168, 174
  types of 2
pens; "unforgiving" aspect, felt-tipped 12
penumbrae 50, 54, 57
photosphere 49–57
  faculae 54, 57
  granulation 53
  limb darkening 53, 54
  penumbrae 54, 57
  sunspots 50, 54, 57
  umbrae 50, 54, 57
  white light 50, 52–54, 57
plage 57, 65–67
planetary nebulae 141–145
  central star 141
  details 142
  field stars 141–142
  framing 141
  NGC 2392 141–145
  observing 141
  shading 142
Polar Cap 86, 89

polar region
  Jupiter 80
  Saturn 76
position angle, solar prominences 61
positive sketching, deep sky 159–161
practice (see Carnegie Hall)
preceding limb, planetary 69
projection solar sketching 52–57
prominences 58–63, 67
  black on white 62
  color on black 61, 67
  white on black 58–61
pseudonucleus, comet 28–30, 34

R
record keeping 40–42
Rite in the Rain 6, 53
Rukl Atlas 2

S
sanding pad 54, 127, 154
Saturn 72
scanning 159
seeing conditions 69
shading, planetary sketching 86
shading, sequence 110
shadows, Moon 3
  importance of early development 6
sharpening
  blending stump 6, 54, 154
  pencil 126–127
sketch
  contour 145–149
  improve skill 10, 12
  naked eye 145–149
sketch board 17
sketch lighting, (see light source) 42–45
sketching tools, planetary 70, 71
solar (see Sun)
solar phenomena 47, 65, 68
Sombrero Galaxy 174
star (see Sun)
star clusters (see open clusters and globular clusters) 97–125
Star sketching 126–127

stars, field stars/framework 25–26, 32–33, 100–101, 106, 108–109, 114, 118, 135–136, 141–142
stars, galaxy 168, 175
stellar geometry 101–102
stellar magnitude, conveying 30, 102–103, 111–113, 116, 124, 126, 138, 143, 146, 153
stippling 12, 13, 122–124, 129–131, 168
Strathmore
  lunar 16
  solar 58–67
strip rendering 82
Sun 47–68
sunspot 50, 54, 57, 67
sunspot recording 50

T
tail, comet 26–28, 35–37
template circle 40–41
template preparation 32–33, 145–146
templates 40–41
  deep sky 181
  Jupiter 180
  Lunar 178
  Saturn 179
  star atlas 32–33, 145–146
terminator, lunar 1
tones, creating gray lunar tones 14
tortillon 154
tracing 32–33, 145–146, 159–161
Trifid Nebula 135–141

U
umbrae 50, 54, 57
unresolved stars/granularity 109–111, 115–116, 119–125, 129–131
Ursa Major 171

W
waning Moon 1
warnings, solar precautions 49–50, 52–53
waxing Moon 1
white light solar filter 49–50, 52–54, 57
white on black deep sky sketching 159–161
white ovals 84